AI FOOD SERVICE

Sarah R. Labensky

Prentice Hall, Upper Saddle River, New Jersey 07458

Library of Congress Cataloging-in-Publication Data

Labensky, Sarah R.
   Applied math for food service / Sarah R. Labensky
   Includes index.
   ISBN 0-13-849217-4
   1. Food service--Mathematics. I. Title.
TX911.3.M33L33   1998
   647.96'0151--dc21   97-22923
                                                                                       CIP

Acquisitions editor: *Neil Marquardt*
Editorial production supervision: *Barbara Marttine Cappuccio*
Director of manufacturing and production: *Bruce Johnson*
Managing editor: *Mary Carnis*
Manufacturing buyer: *Ed O'Dougherty*
Cover design: *Joe Sengotta*
Cover illustration: *Coco Musada*
Creative director: *Marianne Frasco*
Marketing manager: *Frank Mortimer, Jr.*

© 1998 by Prentice-Hall, Inc.
Upper Saddle River, New Jersey 07458

All rights reserved. No part of this book may be
reproduced, in any form or by any means,
without permission in writing from the publisher.

Printed in the United States of America

16 17 18 19 20

ISBN 0-13-849217-4

Prentice-Hall International (UK) Limited, London
Prentice-Hall of Australia Pty. Limited, Sydney
Prentice-Hall Canada Inc., Toronto
Prentice-Hall Hispanoamericana, S.A., Mexico
Prentice-Hall of India Private Limited, New Delhi
Prentice-Hall of Japan, Inc., Tokyo
Pearson Education Asia Pte. Ltd., Singapore
Editora Prentice-Hall do Brasil, Ltda., Rio de Janeiro

CONTENTS

ACKNOWLEDGMENTS vii

INTRODUCTION 1

CHAPTER 1: MEASUREMENTS AND CONVERSIONS 3
    Measurement Formats 4
    Measurement Systems 7
    Converting Grams and Ounces 9
    Review Questions 17

CHAPTER 2: RECIPE CONVERSIONS 19
    Portion Size and Production Quantity 20
    Recipe Conversions 22
    Converting Total Yield 23
    Converting Portion Size 26
    Additional Conversion Problems 30
    Review Questions 33

CHAPTER 3: UNIT AND RECIPE COSTING 35
    Unit Cost 36
    Recipe Cost 45
    Labor Cost 48
    Review Questions 51

CHAPTER 4: YIELD TESTS 53
    Ingredient Yields 54
    Raw Yield Tests Without By-Products 55

CONTENTS

      Applying Yield Factors   60
      Raw Yield Tests With By-Products   65
      Cooked Yield Tests   69
      Using Yield Tests for Purchasing   72
      Making Purchasing Decisions   74
      Review Questions   75

CHAPTER 5: INVENTORY AND FOOD COST PERCENTAGES   77
      Inventory   78
      Shrinkage   81
      Cost of Goods Sold   83
      Food Cost Percentages   86
      Review Questions   91

CHAPTER 6: CONTROLLING FOOD COSTS   93
      Review Questions   103

CHAPTER 7: MENU PRICING   107
      Cost-Based Menu Pricing   108
      Profit-Based Menu Pricing   114
      Non-Cost-Based Pricing   116
      Psychological Impact of Pricing   117
      Review Questions   119

CONTENTS

APPENDIX 121

    Common Container Sizes   122

    Tablespoons per Ounce   124

    Recipe Costing Form   125

    Yield Testing Form   126

    Cooked Yield Test Form   127

    Daily Production Report   128

GLOSSARY 129

INDEX 137

ACKNOWLEDGMENTS

I wish to thank all of my students, past, present, and future, for helping to create this book. Without their insightful questions and their desire to learn, this text would not have been necessary. I am also grateful to Steve Labensky for his editorial assistance and to Neil Marquardt, my editor at Prentice Hall, for supporting this project.

Most of all I wish to thank Dave and all the other students who understood and said "thank you."

INTRODUCTION

Today's professional chef must master more than the basics of béchamel, butchering, and bread-baking. He or she must be equally skilled in the business of food service.

Although computers are almost as common in kitchens as whisks, no machine can ever replace a chef's hands-on ability to apply simple mathematics to real-world situations. Accurate measurements, portion control, and proper food handling directly effect the food service operation's bottom line. In addition, modern chefs must, at the very least, be able to conduct yield tests, calculate recipe costs, and use food cost percentages.

This handbook acquaints culinary students with basic mathematics and procedures for controlling food costs. It does not attempt to cover accounting principles or computer application. It will, however, provide a foundation of practical techniques helpful for anyone willing to use common sense and a calculator.

CHAPTER 1

MEASUREMENTS AND CONVERSIONS

LEARNING OBJECTIVES:

- Understand measurement formats and systems
- Understand conversions to and from the metric system
- Learn basic measurement equivalents

CHAPTER 1

Measurement Formats

Accurate measurements are one of the most important aspects of food production. Ingredients and portions must be correctly measured to ensure consistent product quality. In other words, the chef must be able to prepare a recipe the same way each time and portion sizes must be the same from one order to the next.

In a kitchen, measurements may be made in three ways: weight, volume, and count.

Weight refers to the mass or heaviness of a substance. It is expressed in terms such as pounds, ounces, grams, and tons. Weight may be used to measure liquid or dry ingredients (for example, 2 pounds of eggs for a bread recipe) and to measure portions (for example, 4 ounces of sliced turkey for a sandwich). Because weight is generally the most accurate form of measurement, portion scales or balance scales are commonly used in kitchens.

Volume refers to the space occupied by a substance. This is mathematically calculated as height × width × length. It is expressed in terms such as cups, quarts, gallons, teaspoons, fluid ounces, bushels, and liters. Volume is most commonly used to measure liquids (for example, 2 quarts of stock). It may also be used for dry ingredients when the amount is too small to be accurately weighed (for example, 1/4 teaspoon of salt). Although measuring by volume is somewhat less

Measurements and Conversions

accurate than measuring by weight, it is generally quicker to measure by volume.

Frequently, mistakes are made in food preparation by chefs who wrongly assume that weight and volume are equal. Do not be fooled! **One cup does not always weigh 8 ounces.**

The word ounce has two meanings: it is a measure of weight and a measure of volume (more accurately known as a fluid ounce). While it is true that one standard cup contains 8 *fluid* ounces, it is not true that the contents of that standard cup will *weigh* 8 ounces. For example, the total weight of the diced apple pieces that fill a measuring cup will vary depending on the size of those pieces. Errors are commonly made in the bakeshop by chefs who assume that 8 ounces of flour is the same as one cup of flour. In fact, one cup of flour weighs only about 4-1/2 ounces.

> ! **Weight** refers to the heaviness of a substance.
>
> **Volume** refers to the space occupied by that substance.

CHAPTER 1

It is not unusual to see both weight and volume measurements used in a single recipe. **When a recipe ingredient is expressed in weight, weigh it. When it is expressed as a volume, measure it**. For example, if a recipe calls for 2 cups of whole wheat flour you should measure that flour with a dry measuring cup. It is not necessary to convert the flour to a weight measurement in order to prepare the recipe.

Like most rules, however, this one has exceptions. The weight and volume of water, butter, eggs, and milk are, in each case, essentially the same. For these ingredients you should use whichever measurement is most accurate and convenient.

> **?** A recipe calls for 8 tablespoons of cold butter. How can you measure this butter easily and accurately?

Count refers to the number of individual items. Count is used in recipes (for example, 4 eggs) and in portion control (for example, 2 fish fillets). Count is also commonly used in purchasing to indicate the size of individual items. For example, a "96 count" case of

Measurements and Conversions

lemons means that a 40-pound case contains 96 individual lemons; a "115-count" case means that the same 40 pound case contains 115 individual lemons. So, the lemons in the 96-count case are larger than those in the 115-count case. The higher the count number, the smaller each individual item will be.

Shrimp is another item commonly sold by count. One pound of shrimp may contain from 8 to 100 or more shrimp depending on the size of the individual pieces. When placing an order, the chef must specify the desired count. For example, when ordering one pound of 21/25 count shrimp the chef expects to receive not less than 21 pieces of shrimp nor more than 25 pieces.

> **?** You order 5 pounds of "16-count" fresh Fanny Bay oysters. How many individual oysters do you expect to receive?

Measurement Systems

The measurement formats of weight, volume, and count are used in the imperial, U.S., and metric measurement systems. Because each of these systems is

7

CHAPTER 1

used in modern food service operations, you should be able to prepare recipes written in any of the three.

The **U.S. SYSTEM**, with which you are probably familiar, is the most difficult system to understand. It uses pounds for weight and cups for volume. For specific equivalents see Chart 1.2, found on page 14.

The **IMPERIAL SYSTEM** is used in Great Britain, Canada, and a few other countries. It uses pounds and ounces for weight and pints and fluid ounces for volume. Imperial volumes are slightly larger than U.S. volumes, however. For example, an imperial gallon is equal to 4.546 liters or about 1-1/5 U.S. gallons.

The **METRIC SYSTEM** is the most commonly used system in the world. Developed in France during the late 18th century, it was intended to fill the need for a mathematically rational and uniform system of measurements. The metric system is a decimal system in which the *gram*, the *liter,* and the *meter* are the basic units of weight, volume, and length, respectively. Larger or smaller units of weight, volume, and length are formed by adding a prefix to the words gram, liter, or meter. Some of the more commonly used prefixes in food service operations are deca- (10), hecto- (100), kilo- (1000) and milli- (1/1000). Thus, a kilogram is 1000 grams; a decameter is 10 meters and a milliliter is 1/1000 of a liter. The metric temperature scale, known as Celsius or centigrade, is based on 100 degrees. 0° C

Measurements and Conversions

is the freezing point of water (32° F) and 100° C is the boiling point of water (212° F).

The most important thing to know about using the metric system is that, in general, **you should not convert between the metric system and the U.S. system in recipe preparation**. If a recipe is written in metric units, use metric measuring equipment; if it is written in U.S. units, use U.S. measuring equipment. Luckily, most modern measuring equipment is marked in both U.S. and metric increments. The need to convert amounts from one system to the other will only arise if the proper equipment is unavailable.

Like most rules, this one also has an exception, which most often arises in the context of food costing. Imported foodstuffs may be labeled by metric weight or volume only. If that ingredient is then used in a U.S. measurement, it will be necessary to convert for costing purposes. For example, a pastry chef purchases French praline paste by the kilogram but uses it by the ounce. To accurately determine the cost of a recipe containing 3 ounces of praline paste the kilogram price must be converted to an ounce price.

Converting Grams and Ounces

Because it will sometimes be necessary to convert weights or measurements from the metric system to the U.S. system, or vice-versa, a table of

CHAPTER 1

precise equivalents is provided in Chart 1.3, found on page 15. Perhaps the most commonly used conversions are simple ones, however—from grams to ounces or ounces to grams.

As you can see from Chart 1.3, 1 ounce equals 28.35 grams. This is often rounded to 28 for convenience. So, to convert ounces to grams **multiply** the number of ounces by 28.

$$8 \text{ ounces} \times 28 \text{ grams} = 224 \text{ grams}$$

And to convert grams to ounces **divide** the number of grams by 28.

$$\frac{224 \text{ grams}}{28 \text{ grams}} = 8 \text{ ounces}$$

# of ounces × 28 = # of grams

# of grams ÷ 28 = # of ounces

Use 28.35

Measurements and Conversions

Exercise 1.1

Practice converting grams to ounces by completing the following:

4 oz. = __113.4__ grams

1 lb. = __16__ ounces = __453.6__ grams

224 g = __7.9__ ounces

1 kg = __1000__ grams = __35.3__ ounces

The next exercise refers back to the pastry chef's dilemma on page 9.

Exercise 1.2

One kilogram of praline paste costs $28.00. The recipe used 3 oz. of praline. What is the cost for that 3 oz.? (Note: The correct answer can be determined using several different approaches.)   2.34

The following list is presented to help you develop your judgment skills when preparing new recipes or making conversions. It is not a substitute for accurate conversions, but it can help you visualize the size of a metric quantity, which is important in selecting pots and pans, mixing bowls, and utensils.

CHAPTER 1

size of a metric quantity, which is important in selecting pots and pans, mixing bowls, and utensils.

- A *kilogram* is about 2.2 pounds.
- A *gram* is about 1/30 ounce.
- A pound is about 450 *grams*. 453.6
- A *liter* is slightly more than a quart.
- A *centimeter* is slightly less than 1/2 inch.
- 0° *Celsius* is the freezing point of water (32°F).
- 100° *Celsius* is the boiling point of water (212°F).

Because the metric system is based on multiples of 10, it is extremely easy to increase or decrease recipe amounts. Exercise 1.3 demonstrates how easy it is to change quantities in a metric recipe.

Exercise 1.3

Divide the following recipe in half:

Flour	1 kilogram	500	g
Sugar	4 grams	2	g
Water	1 liter	500	ml
Eggs	2 whole	1	each
Butter	250 grams	125	g
Vanilla	10 ml	5	ml

Measurements and Conversions

Charts 1.1 and 1.2 provide information on measurement abbreviations and equivalents. (Additional measurement charts are provided in the Appendix.) By memorizing key facts you will be able to figure out unknown facts for yourself as you work with recipes. Although it may seem hard at first, there is no substitute for knowing this information. In fact, the more you measure ingredients and portion foods, the sooner this information will become second nature to you.

Chart 1.1

ABBREVIATIONS

t. *or* tsp.	=	teaspoon
T. *or* Tbsp.	=	tablespoon
c.	=	cup
pt.	=	pint
qt.	=	quart
No. *or* #	=	number
lb. *or* #	=	pound
g	=	gram
kg	=	kilogram
ml	=	milliliter
l *or* lt	=	liter
F	=	Fahrenheit
C	=	Celsius

CHAPTER 1

Chart 1.2

COMMON EQUIVALENTS IN THE U.S. SYSTEM

1/8 teaspoon	=	dash
1 teaspoon	=	1/3 tablespoon *or* 60 drops
▶ **3 teaspoons**	=	**1 tablespoon *or* 1/2 fl. oz.**
1/2 tablespoon	=	1-1/2 teaspoons
1 tablespoon	=	3 teaspoons or 1/2 fl. oz.
▶ **2 tablespoons**	=	**1 fluid ounce**
3 tablespoons	=	1-1/2 fl. oz. *or* 1 jigger
4 tablespoons	=	1/4 cup *or* 2 fl. oz.
8 tablespoons	=	1/2 cup *or* 4 fl. oz.
12 tablespoons	=	3/4 cup *or* 6 fl. oz.
▶ **16 tablespoons**	=	**1 cup *or* 8 fl. oz.**
1/8 cup	=	2 tablespoons *or* 1 fl. oz.
▶ **1/4 cup**	=	**4 tablespoons *or* 2 fl. oz.**
1/3 cup	=	5 tablespoons + 1 teaspoon
3/8 cup	=	1/4 cup + 2 tablespoons
▶ **1/2 cup**	=	**8 tablespoons *or* 4 fl. oz.**
5/8 cup	=	1/2 cup + 2 tablespoons
3/4 cup	=	12 tablespoons *or* 6 fl. oz.
7/8 cup	=	3/4 cup + 2 tablespoons
▶ **1 cup**	=	**16 tablespoons *or* 1/2 pint *or* 8 fl. oz.**
▶ **2 cups**	=	**1 pint *or* 16 fl. oz.**
1 pint	=	2 cups *or* 16 fl. oz.
▶ **1 quart**	=	**2 pt. *or* 4 cups *or* 32 fl. oz.**
▶ **1 gallon**	=	**4 qt. *or* 8 pt. *or* 16 cups *or* 128 fl. oz.**
2 gallons	=	1 peck
4 pecks	=	1 bushel

1 Liter = 33.8 oz

Chart 1.3
PRECISE METRIC EQUIVALENTS

Volume:

1/4 teaspoon	=	1.23 milliliters
1/2 teaspoon	=	2.46 milliliters
1 teaspoon	=	**4.93 milliliters**
1-1/4 teaspoons	=	6.16 milliliters
1-1/2 teaspoons	=	7.39 milliliters
1-3/4 teaspoons	=	8.63 milliliters
2 teaspoons	=	9.86 milliliters
1 tablespoon	=	14.79 milliliters
1 fluid ounce	=	29.57 milliliters
2 tablespoons	=	29.57 milliliters
1/4 cup	=	59.15 milliliters
1/2 cup	=	118.3 milliliters
1 cup	=	**236.59 milliliters**
2 cups *or* 1 pt.	=	473.18 milliliters
3 cups	=	709.77 milliliters
4 cups *or* 1 qt.	=	946.36 milliliters
4 qt. *or* 1 gallon	=	3.785 liters

Weight:

1 ounce	=	**28.35 grams**
8 ounces *or* 1/2 pound		
16 ounces *or*	=	226.8 grams
1 pound	=	**453.6 grams**
2 pounds	=	907.2 grams
2 pounds + 3 ounces	=	1 kilogram *or* 1000 grams

Measurements and Conversions

Review Questions for Chapter 1

Define the following terms:

GALLON –

WEIGHT – The heaviness of a substance

METRIC SYSTEM – The global language of measurement and has been adopted by almost every country. Based on the number 10

1. Milk costs $1.97 per gallon. Your bread recipe uses 6 ounces. How much does that amount cost?
   0.015 $.09

2. A soup recipe uses 6 quarts of milk. How much does that amount cost? $2.96

3. How many teaspoons are there in one cup?
   48 tsp.

4. How many pounds are there in one gallon of water?
   8 #

5. How many ounces of yeast are in an envelope that contains 7 grams?
   1/4 oz

CHAPTER 2

RECIPE CONVERSIONS

LEARNING OBJECTIVES:

- Understand how to determine quantities needed

- Understand how to convert recipe yields

- Understand the role of judgment and skill in recipe conversions

CHAPTER 2

Portion Size and Production Quantity

An often overlooked or underappreciated aspect of commercial food preparation is the initial determination of portion size. Without determining an appropriate portion size (and then applying that to the anticipated customer count), the chef might prepare too much food, or not prepare enough food — both of which are costly, undesirable mistakes. There are no hard and fast rules for these calculations, however. Each situation depends on the chef's or manager's judgment and experience.

A proper portion size is one that has been determined following a thoughtful analysis of several factors: the type of operation or event, customer satisfaction, competition, food costs, and merchandising concerns. Each operation must determine its own portion standards, usually based upon a blend of quantity, cost, and merchandising issues.

Quantity Determine the amount of each food item that the average patron would prefer to consume given the time of day or meal, the event, and the total selection offered. For example, fettuccine Alfredo can be portioned as a 4-ounce serving for an appetizer or a 7-ounce serving for an entrée.

Cost Determine the amount of food that can be provided for the price charged, while still covering expenses and making a reasonable profit. For example,

Recipe Conversions

if a shrimp cocktail sells for $9.95, you can determine how many shrimp to serve by determining how much of that price can be "spent" on the cost of raw ingredients.

Merchandising Determine the amount of each portion by evaluating what will look best, attract attention, or fit on or in available serviceware. For example, if the only soup bowls available hold 5 ounces, it is not necessary to prepare 7 ounces of soup per person.

Once the appropriate portion size is determined, that amount is multiplied by the customer count in order to calculate the total amount to produce.

Portion Size × *Number of Servings Needed*

= *Total to Produce*

Exercise 2.1

A catering company is preparing 75 boxed lunches, which will each contain a 4-ounce serving of fruit salad. Another event scheduled for the same day includes fruit salad on a buffet designed to serve 150 guests. In total, how much fruit salad would you anticipate that the caterer should prepare? Why?

CHAPTER 2

Recipe Conversions

Whether 6 servings or 60, every recipe is designed to produce a specific amount of product, known as *yield*. A recipe's yield may be expressed in *volume, weight,* or *servings* (examples: 1 quart of sauce; 8 pounds of bread dough; 8 half-cup servings). If the expected yield does not meet your needs, you must convert (i.e., increase or decrease) the recipe by converting the amount of each ingredient in the recipe.

Recipe conversion is sometimes complicated by *portion size conversions.* For example, it may be necessary to convert a recipe that initially produces 24 8-ounce servings of soup into a recipe that produces 62 6-ounce servings.

Cutting a recipe in half or doubling a recipe is done almost instinctively. What happens mathematically when you do this?

▶ To "double a recipe," the amount of each ingredient is actually multiplied by 2.

▶ To "cut a recipe in half," the amount of each ingredient is actually multiplied by .5.

Although not all recipe conversions use such nice even numbers, it is just as easy to change recipe yields by uneven amounts as it is to double or halve recipes. The mathematical principle is the same: **Each ingredient is multiplied by a conversion factor.**

Recipe Conversions

Do not take shortcuts by estimating recipe amounts or conversion factors. Inaccurate conversions lead to inedible foods, embarrassing shortages, or wasteful excess. Take the time to learn and apply proper conversion techniques.

Converting Total Yield

When portion size remains the same or is unimportant to your calculations, recipe yield is converted by a simple two-step process:

STEP 1: Divide the desired (new) yield by the recipe (old) yield to obtain the conversion factor.

$$\frac{New\ Yield}{Old\ Yield} = Conversion\ Factor$$

STEP 2: Multiply each ingredient quantity by the conversion factor to obtain the new quantity.

Old Quantity × Conversion Factor = New Quantity

CHAPTER 2

Example 2.1

You need to convert a recipe for cauliflower soup. The present recipe yields 1-1/2 gallons. You only need to make 3/4 of a gallon.

First, determine the conversion factor:

$$\frac{.75 \text{ gallon}}{1.5 \text{ gallons}} = .50$$

(Need help in changing fractions to decimal numbers? Please see your instructor for guidance.)

The conversion factor (CF) is applied to each ingredient in the soup recipe:

Cauliflower Soup				
	Old quantity	× CF		= New quantity
Cauliflower, chopped	5 lb.	× .5	=	2-1/2 lb.
Celery stalks	4	× .5	=	2
Onion	1	× .5	=	1/2
Chicken stock	2 qt.	× .5	=	1 qt.
Heavy cream	6 cups	× .5	=	3 cups

Recipe Conversions

When calculating a conversion factor, the new and old yields can be in any unit (cups, ounces, pounds, servings, and so on) **as long as** the numerator and the denominator are in the same unit.

For example, in the cauliflower soup exercise, the same conversion factor can be obtained after first converting the recipe amounts to fluid ounces:

$$\frac{96 \text{ fl. oz.}}{192 \text{ fl. oz.}} = .50$$

(Remember, 1 gallon contains 128 fluid ounces!)

$$\frac{\textit{New Yield}}{\textit{Old Yield}} = \textit{Conversion Factor}$$

Exercise 2.2

A recipe you normally use produces 6 lb. of chocolate chip cookie dough, but you need 10 lb. of dough. What is the correct conversion factor?

CHAPTER 2

Converting Portion Size

A few additional steps are necessary to convert recipes when portion sizes must also be changed.

STEP 1: Determine the total yield of the existing recipe by multiplying the number of portions by the portion size.

Original Portions × *Original Portion Size* = *Total Old Yield*

STEP 2: Determine the total yield desired by multiplying the new number of portions by the new portion size.

Desired Portions × *Desired Portion Size* = *Total New Yield*

STEP 3: Obtain the conversion factor by dividing the new total yield by the old total yield.

$$\frac{\textit{Total Yield (New)}}{\textit{Total Yield (Old)}} = \textit{Conversion Factor}$$

STEP 4: Multiply each ingredient quantity by the conversion factor.

Old Quantity × *Conversion Factor* = *New Quantity*

Recipe Conversions

To summarize, the complete formula looks like this:

$$\frac{\textit{Original Portions} \times \textit{Original Portion Size} = \textit{Total Yield (New)}}{\textit{Desired Portions} \times \textit{Desired Portion Size} = \textit{Total Yield (Old)}} = CF$$

$$\textit{Old Quantity} \times \textit{Conversion Factor (CF)} = \textit{New Quantity}$$

Example 2.2

The original cauliflower soup recipe produced 1-1/2 gallons, which is the same as 48 4-ounce servings. Now you need 72 6-ounce servings of the soup for a banquet. The following steps show how this conversion is made.

CHAPTER 2

STEP 1: Total original yield is 48 × 4 oz. = 192 oz.

STEP 2: Desired yield is 72 × 6 oz. = 432 oz.

STEP 3: Conversion factor is calculated by dividing desired (new) yield by original (old) yield: 432 ÷ 192 = 2.25.

STEP 4: Old ingredient quantities are multiplied by conversion factor to determine new quantities:

Cauliflower Soup

	Old quantity	×	CF	=	New quantity
Cauliflower, chopped	5 lb.	×	2.25	=	11-1/4 lb.
Celery stalks	4	×	2.25	=	9 lb.
Onion	1	×	2.25	=	2-1/4
Chicken stock	2 qt.	×	2.25	=	4-1/2 qt.
Heavy cream	6 cups	×	2.25	=	13-1/2 cups

Recipe Conversions

When a recipe's yield is increased, it sometimes becomes impractical to measure the new, larger quantities by cup, tablespoon, or the like. Your knowledge of measurements will then be helpful in changing recipe amounts into more manageable units. For example, the 4-1/2 quarts of chicken stock in the Cauliflower Soup recipe above could be *physically* measured as 1 gallon plus 1 pint.

> **?** The new recipe for Cauliflower Soup requires 13.5 cups of heavy cream. What is the most convenient way to measure this amount of cream?

Exercise 2.3

Your recipe for Devil's Food Cake makes 2 gallons of batter. You need to make 100 cupcakes. Each cupcake tin holds 4 fluid ounces of batter. Find the appropriate conversion factor.

The only way to learn conversion techniques is to **practice, practice, practice.** The review questions for this section are an opportunity to practice a variety of measurement and conversion skills.

CHAPTER 2

Additional Conversion Problems

When making very large recipe changes—for example, from five to 500 portions or 600 to 36 portions—you may encounter additional problems. The mathematical conversions described above do not take into account changes in equipment, evaporation rates, cooking times, or unforeseen recipe errors. Good cooks learn to use their judgment, knowledge of cooking principles, and skills to compensate for these factors.

Equipment When you change the size of a recipe, you must often change the equipment used as well. Problems arise, however, when the production techniques previously used no longer work with the new quantity of ingredients. For example, if you normally make a muffin recipe in small quantities by hand and you increase the recipe size, it may be necessary to prepare the batter in a mixer. If mixing time remains the same, the batter may become overmixed, resulting in poor quality muffins. Trying to prepare a small amount of product in equipment that is too large for the task will also affect its quality.

Evaporation Equipment changes can also affect product quality because of changes in evaporation rates. Increasing a soup recipe may require substituting a tilt skillet for a saucepan. But because a tilt skillet provides more surface area for evaporation than does a saucepan, reduction time must be decreased to prevent

Recipe Conversions

over-thickening the soup. The increased evaporation caused by an increased surface area may also alter the strength of seasonings.

Recipe Errors A recipe may contain errors in ingredients or techniques that are not obvious when prepared in small quantities. When increased, however, small mistakes often become big (and obvious) mistakes and the final product suffers. The only solution is to test recipes carefully and rely on your knowledge of sound cooking principles to compensate for unexpected problems.

Note that herbs and seasonings such as salt are rarely multiplied by the mathematical conversion factor when enlarging a recipe. In other words, if you increase a recipe eight times you **do not** automatically increase the thyme and salt in that recipe eight times. You should rely instead on experience and taste to determine how much of each seasoning to add.

Time Do not multiply time specifications given in a recipe by the conversion factor used with the recipe's ingredients. All things being equal, *cooking time* will not change when recipe size or yield is changed. For example, a muffin requires the same amount of baking time whether you are preparing one dozen or 14 dozen. Cooking time will be affected, however, by changes in evaporation rate or heat conduction caused by changes in pots and pans or bakeware.

CHAPTER 2

Mixing time may change when recipe size is changed. Different equipment may perform mixing tasks more or less efficiently than the equipment previously used. Again, rely on experience and good judgment.

Exercise 2.4

You have prepared a standard muffin recipe several times with no problems. The chef asks you to make four times as many muffins. You convert the recipe properly but the batter seems much thicker than normal and the mixing bowl you always use isn't large enough. The baked muffins are unacceptable and cannot be served. What might have gone wrong?

Recipe Conversions

Review Questions for Chapter 2

Convert the following recipes as indicated:

1. Beef Tenderloin with Mushrooms

Portions: 8   *2.25*
Portion size: 8 oz.

Butter	2 oz.	— 4.5
Onions	4 oz.	— 9
Flour	1 Tbsp.	— 2.25
Mushrooms	1/2 lb.	—
Tenderloin	2-1/2 lb.	— 5.625
White Wine	1/2 cup	1.125
Mustard	2 tsp.	— 4.5
Brown Sauce	1-1/2 pt.	— 3.375
Cream	1 cup	— 2.25
Salt & Pepper	to taste	

Convert recipe to yield 18 8-ounce serving.

2. Baked Summer Squash

Yield: 4 servings

Squash	3 cups	— 35.25 *47*
Milk	1/4 cup	— 2.9375 = 11.75
Butter	2 Tbsp.	— 23.5 *4*
Salt	1 tsp.	— 11.75
Paprika	1/4 tsp.	— 2.9375

Convert recipe to yield 47 servings.

CHAPTER 2

3. **Vanilla Sauce**

   Yield: 2-1/2 pt.
Egg Yolks	12
Sugar	8 oz.
Milk	1 qt.
Vanilla	1 Tbsp.

   Convert recipe to yield 2-1/2 qt.

4. **Soft Dinner Rolls**

   Portions: 64
   Portion size: 1.25 oz.
Water	600 ml.
Yeast	60 g
Flour	1300 g
Salt	30 g
Sugar	120 g
Milk Powder	60 g
Shortening	50 g
Butter	70 g
Eggs	120 ml.

   $3.5 = \frac{280}{80}$

   $3.125 = \frac{25}{40}$

   A. Convert recipe to yield 140 2-ounce rolls. — 3.5
   B. Convert recipe to yield 20 1-1/4 ounce rolls. — 3.125

5. List and explain three things that can affect successful recipe size changes.

34

CHAPTER 3

UNIT AND RECIPE COSTING

LEARNING OBJECTIVES:

- Understand how to calculate unit cost

- Understand and use typical invoices

- Understand how to calculate the cost of a recipe

- Understand how to calculate portion cost

CHAPTER 3

Unit Cost

Food service operations purchase most food items from suppliers in bulk or wholesale packages. For example, canned goods are purchased by the case; produce by the flat, case, or lug; and flour and sugar by 50- or 100-pound bags. Even fish and meats are usually purchased in large cuts, not individual serving-sized portions. The purchased amount is rarely used for a single recipe, however. It must be broken down into smaller units such as pounds, cups, quarts, or ounces.

In order to allocate the proper ingredient costs to the recipe being prepared, it is necessary to convert *as-purchased costs* or prices to *unit costs* or prices. To find the unit cost (i.e., a single egg) in a package containing multiple units (i.e., a 30-dozen case), *divide* the as-purchased (A.P.) cost of the package by the number of units in the package.

$$\frac{A.P.\ Cost}{Number\ of\ Units} = Cost\ per\ Unit$$

Unit and Recipe Costing

Example 3.1

A case of #10 cans contains 6 individual cans. If a case of tomato paste costs $23.50, then each can costs $3.92.

$$\frac{\$23.50}{6} = \$3.92$$

If your recipe uses less than the total can, you must continue dividing the cost of the can until you arrive at the appropriate unit amount. Continuing with the tomato paste example, if you need only 1 cup of tomato paste, divide the can price ($3.92) by the total number of cups contained in the can to arrive at the cost per cup (unit). *(The list of canned good sizes in the Appendix shows that a #10 can contains approximately 13 cups.)* Using the formula, we find that each cup costs $0.30.

$$\frac{\$3.92}{13} = .30$$

The cost of one cup can be reduced further if necessary. If the recipe uses only 2 tablespoons of tomato paste, divide the cost per cup by the number of tablespoons in a cup. *(Chart 1.2 shows that there are 16 Tbsp. per cup.)* As you can see, the final cost for 2 tablespoons of this tomato paste is $0.037.

CHAPTER 3

$$\frac{.30}{16} = .018 \times 2 = \$0.037$$

The following exercise will help you understand unit costing procedures and provide additional practice with measurement equivalents.

> **!** UNIT COST = The price paid to acquire **one** of the specified items.

Unit and Recipe Costing

Exercise 3.1

Find the unit cost for each item:

Item	A.P. package	Cost	Unit Cost
Flour, cake	50# bag	11.20	.224 lb.
Beans, fresh	28# bushel	20.72	.74 lb.
Tomato puree	#10 can	3.10	.2384 cup
Milk	gallon	2.30	.0179 oz.
Corn syrup	gallon	9.45	.5906 cup
Salt	50# bag	4.40	4.40/50 = .088 .088÷16 = .0055÷2 = .00275 Tbsp.
Leeks	crate/12 bunches	17.40	1.45 bun
Onions	25# bag	14.00	.56 lb.
Butter	36# case	42.70	.0741 oz.
Cauliflower	flat/9 heads	16.25	1.80555 ea.

39

CHAPTER 3

Cost information is usually provided to the chef or manager on purchase invoices, such as the ones shown on the following pages. It may also be necessary to examine a product's label or package to determine some information such as size or weight.

Exercise 3.2

Examine the following invoices to determine the unit cost for these items:

Item	Price		Unit Cost	
Potatoes	9.05	$ 50	.181	per lb.
Parsley	9.25	$	.1542	per bun
Apples	1540	$ 88	.175	per each
King Crab Legs	9.40	$		per each
Wine Vinegar	14.95	$	.29609	per cup
Butter	1.15 1#	$	.071875	per oz.
WOG Fryers	1.77	$	1.77	per lb.
Sugar	25.99	$	.5198	per lb.
Tea	19.95	$	.09975	per serving

40

Stern produce
"Fresh Fruits and Vegetables Since 1917"

3200 South 7th Street
Phoenix, Arizona 85040
602.268.6628
fax 602.268.2696

INVOICE

TRIP STOP 2-09-140
INVOICE NO. 142616
SLSM 6
PAGE 1

BILL TO
MARICOPA COMM. COLLEGE
2411 W. 14TH ST
TEMPE AZ 85281-6941

SHIP TO
SCOTTSDALE COMMUNITY COL.
CULINARY ARTS
9000 E. CHAPARRAL
SCOTTSDALE AZ 85256

DATE 06/04/97
ACCOUNT 71800
TELEPHONE 602-423-6242
TERMS/PURCHASE NUMBER NET 30 DAYS

METHOD OF PAYMENT
☐ ON ACCOUNT CK #
☐ COD CK #
☐ CASH AMOUNT

SLOT	K	QUANTITY	UNIT	BRAND	OTH	TOT	WHT	CUBE	DESCRIPTION	SIZE	UNIT	TAX	ITEM NO.	PRICE	AMOUNT
A304A		1.00	CTN						LEEKS	1/12	EA.		10400	10.25	10.25
B109A		1.00	CTN						PARSLEY	1/60	EA.		27700	9.25	9.25
B101A		1.00	CSE						LETTUCE CLEANED @ TRIMMED	4/6	HD		02100	10.45	10.45
D134A		1.00	CSE						POTATO IDAHO BAKER 60 CNT	1/50	LB.		05400	9.05	9.05
A611A		1.00	FLT						TOMATO 5X6X2 *REPACK*	1/18	LB.		08300	12.25	12.25
A101A		1.00	CTN						ORANGE 56 SUNKIST	1/40	LB.		14900	14.95	14.95
A204A		1.00	FLT						MUSHROOM MEDIUM	1/10	LB.		23500	14.95	14.95
B312A		1.00	CTN						APPLE RED DELICIOUS 88 CT	1/40	LB.		31400	15.40	15.40
B419A		1.00	CSE						PEACH	1/22	LB.		35200	19.75	19.75
A104A		1.00	CSE						HONEYDEW 8 CNT	1/8	EA.		37200	9.65	9.65
B304A		1.00	FLT						STRAWBERRIES	1/12	EA.		39300	9.35	9.35
CATEGORY		0-0000-0000-00			OTH	TOT									
PIECES					11	11	0	14			312	312			

INVOICE TOTAL 134.60
TAX .00
PAY THIS AMOUNT 134.60

TOTAL DUE ⇨

CHAPTER 3

2721 WEST WILLETTA
PHOENIX, AZ 85009
602-269-7717

02/26/97   9.8716

SOLD TO:
139
WALKER MARKET
250 SWANSON AVE
LAKE HAVASU, AZ 86403

SHIP TO: SAME

INVOICE
704554

Cust phone# 453-8746
Document: 696259   GC: Shipped: 02/26/97
Sales Rep: HOUSE LH   PO: Due Date: COD-OK TO SIGN

PRODUCT NUMBER	DESCRIPTION	U/M	QUANTITY ORDERED	QUANTITY SHIPPED	UNIT PRICE	EXTENDED AMOUNT
10542	CAB TOP ROUND (INSIDES)	LB	3EA	69.00	1.49	102.81
11729	CHOICE NEW YORK 13/UP 1 X 1	LB	2MC	160.00	3.75	600.00
11933	5/UP PSMO TENDERLOIN	LB	4EA	22.00	7.35	161.70
13514	GROUND BEEF-REGULAR BULK {80/20}	LB	2MC	40.00	1.39	55.60
20204	PORK LOIN 13/18 *FRESH* EXCEL #411144	LB	2EA	64.00	1.69	108.16
22630	CITY MEAT LAYOUT BACON 18/22 HICKORY SMOKED 15# MC	LB	1MC	15.00	2.00	30.00
25032	SMOKED PORK HOCKS FLETCHER/FARMER JOHN	LB	1MC	30.00	1.14	34.20
30236	CERT. LAMB RACK FRENCH/SEND 9 BONE-CITY MEAT PRODUCTION	LB	3EA	30.00	6.85	205.50
31146	VEAL LEGS - CUT AND SEND ALL SWISS CLASS AAA	LB	1EA	45.00	3.65	164.25
31523	AAA VEAL RACKS FRESH CUT & SEND SPECIFY CUT SIZE 02	LB	5EA	45.00	5.75	258.75
40707	DUCK 5# UP CASE SALES ONLY WINONA BRAND	LB	1MC	30.00	1.39	41.70
42072	24/8 OZ BNLS CHICKEN BREAST SKIN ON CHEF MAXLOTTE'S	LB	6MC	72.00	1.75	126.00
45155	FRESH WOG FRYERS 3/UP CVP	LB	1MC	73.00	1.77	129.21

NOTICE: Product must be weighed and inspected with driver on delivery. Invoice subject to correction on delivery only. No claims allowed except at time of delivery.
RECEIVED BY

X _____

TOTAL QTY. SHIPPED
CASE COUNT

INVOICE TOTAL

PLEASE PAY THIS AMOUNT
CONTINUED

This transaction is evidenced by purchases listed hereon, constitutes an agreement by buyer and seller and signature of buyer, his agents or his employees shall be acknowledged of same and failure to make payment when due shall be basis for legal action to be taken and buyer agrees to pay all court costs and reasonable attorney fees. It is further agreed and understood at the time of purchase of said goods that any employee or buyer may sign for and receive such goods for buyer, thereby making buyer responsible for this contract in its entirety. It is agreed that time is of the essence of this contract.
NOTICE: PAST DUE ACCOUNTS SUBJECT TO 1½% PER MONTH (18% ANNUALLY), OR THE HIGHEST RATE PERMITTED BY LAW.

ACCOUNTING

Unit and Recipe Costing

2721 WEST WILLETTA
PHOENIX, AZ 85009
602-269-7717

02/26/97   9.8717

| SOLD TO | 139
WALKER MARKET
250 SWANSON AVE
LAKE HAVASU, AZ 86403 | SHIP TO | SAME | | INVOICE
704554 |

Cust phone# 453-8746
Document: 696259   GC: Shipped: 02/26/97
Sales Rep: HOUSE LH   PO: Due Date: COD-OK TO SIGN

PRODUCT NUMBER	DESCRIPTION	U/M	QUANTITY ORDERED	QUANTITY SHIPPED	UNIT PRICE	EXTENDED AMOUNT
	22HD COUNTRY PRIDE#112939					
52175	BUTTER UNSALTED 36/1 RED TAPE OR GRASSLAND	LB	3MC	108.00	1.15	124.20
65109	KING CRAB LEGS 12-14 CT	LB	1MC	20.00	9.80	196.00
66400	CLEAN SQUID 5-8 IN.	LB	1MC	50.00	2.29	114.50

!!!!!!!!!!!!!!!!!!!!!!!!!!!!!
CITY MEAT IS CUSTOMER DRIVEN!
LET US KNOW HOW WE CAN
SERVICE YOU BETTER?

	TOTAL QTY. SHIPPED	873.00	INVOICE TOTAL	2452.58
	CASE COUNT		PLEASE PAY THIS AMOUNT	

NOTICE: Product must be weighed and inspected with driver on delivery. Invoice subject to correction on delivery only. No claims allowed except at time of delivery.
RECEIVED BY

X _____

This transaction is evidenced by purchases listed hereon, constitutes an agreement by buyer and seller and signature of buyer, his agents or his employees shall be acknowledged of same and failure to make payment when due shall be basis for legal action to be taken and buyer agrees to pay all court costs and reasonable attorney fees. It is further agreed and understood at the time of purchase of said goods that any employee or buyer may sign for and receive such goods for buyer, thereby making buyer responsible for this contract in its entirety. It is agreed that time is of the essence of this contract.
NOTICE: PAST DUE ACCOUNTS SUBJECT TO 1½% PER MONTH (18% ANNUALLY), OR THE HIGHEST RATE PERMITTED BY LAW

ACCOUNTING

RYKOFF-SEXTON, INC.

TELEPHONE: (602)352-3300

PLEASE REMIT TO: RYKOFF-SEXTON-PHOENIX
FILE #54806
LOS ANGELES, CA 90074-4806

ENJOY LIFE, EAT OUT MORE OFTEN®

SOLD TO:
STEVE SIMMONS
643 E BARBARITA
GILBERT, AZ 85234

SHIP TO:
SIMMONS CAFE
CULINARY DRIVE
PHOENIX, AZ 85018

TERMS: NET 7 DAYS

SM. NO.	ACCOUNT NO.	INVOICE DATE	PAGE
621	0000124	04/24/97	1

TRUCK	ROUTE	STOP
999		0

INVOICE NO. **114-00411**

CUSTOMER ORDER NO. 0000-0000

QUANT	UNIT	PACK AND SIZE	PRODUCT	DESCRIPTION	WHSE. LOC.	PRICE	EXTENSION	TAX
				INV INVOICE				
				INVOICE SPECIAL INSTRUCTION:				
				DO NOT PICK				
				GIVE INVOICE TO				
				MIKE BYRON				
				XXXXXXXXXXXXXXXXXXXXXX				
				FROZEN & COOLER ITEMS FOLLOW:				
4		1/QUART	74381	PCKRS WHIPPING CREAM STERIL QUART	C0237B	2.65	10.60	
				DRY AND NON FOOD ITEMS FOLLOW:				
2		1/4 OZ	22396	RSCON R/S CONN TARRAGON LEAF 880294	T3323	9.65	19.30	
3		8/25 ENV	02184	CONN 02184/0 CONNOISSEUR TEA ASST 8 PACK	P4321	19.95	59.85	
1		1/50 LB	01161	SER FLOUR ALL PURPOSE SER FAMILY	E4911	13.35	13.35	
1		1/50 LB	22072	C&H GRANULATED SUGAR	E3214	25.99	25.99	
1		4/1 GAL	05592	SER 50638 SER CHAMPAGNE WHITE WINE VINEGAR 50 GRAI	D0121	18.95	18.95	
				ASK YOUR SALES REP ABOUT OUR NEW BOXED BEEF PROGRAM				

QUANT	CUBE	WEIGHT	FOOD ITEMS	PAPER & SANIT.	EQUIPMENT	TOTAL MERCHANDISE		AMOUNT TAXABLE		REMARKS	TAX AMOUNT	TOTAL
12	4	156	148.04			COOLER $ 10.60	FROZEN $	DRY $ 137.44	CASES 8/	0		148.04
				In the event of non-payment on the company's terms, customer agrees to pay reasonable attorney fees and court costs. Returned Checks are subject to a service charge. Past-Due balances are subject to a 1.5% per month service charge.		CASES 4/	CASES 0					(TOTAL PLUS SALES TAX)

DELIVERED BY _____

RECEIVED BY _____

RYKOFF-SEXTON COPY

Unit and Recipe Costing

Recipe Cost

A standard recipe, listing the ingredients and their amounts as well as the number and size of the portions, must be established in order to determine the cost of the completed menu item. Once an accurate recipe is written, the ***Total Recipe Cost*** is calculated with the following two-step procedure:

STEP 1: Determine the cost for the given quantity of each recipe ingredient using the unit costing procedures described previously.

STEP 2: Add all of the ingredient costs together to obtain the total recipe cost.

The total recipe cost can then be broken down into the ***Cost per Portion***, which is the most useful figure for food cost control. To arrive at cost per portion, divide the total recipe cost by the total number of servings or portions produced by that recipe.

$$\frac{\textit{Total Recipe Cost}}{\textit{Number of Portions}} = \textit{Cost per Portion}$$

CHAPTER 3

The *portion* should be the amount that is actually served or sold to customers. In other words, if pecan pie is cut into 8 pieces for service, the recipe cost should be divided by 8 for an accurate portion cost. Dividing the total recipe cost by 6 servings or by 10 servings would provide erroneous information.

> **?** Identify at least two other reasons why it is important to use a standard recipe in a professional kitchen.

The Recipe Costing Form provided in the Appendix is useful for organizing recipe costing information. It provides space for listing each ingredient, the quantity of each ingredient needed, the cost of each unit and the total cost for each ingredient. Portion size and cost per portion are listed at the bottom of the form. There is no space for recipe procedures because they are irrelevant in recipe costing.

When using this costing form, remember that information you write in the **Quantity** column is multiplied by the amount you write in the **Price** column to determine the **Recipe Cost** for each item. The costs of the ingredients are then totaled to determine the **Total Recipe Cost**.

Unit and Recipe Costing

Exercise 3.3 shows the ingredients for an onion soup. Use the unit costs calculated in Exercise 3.1 to complete the recipe costing form. *(Note that the columns for "yield %" and "E.P. price" are not used in this exercise.)*

Exercise 3.3

RECIPE COSTING FORM

Menu Item **Onion Soup**   Date **10-12**
Total Yield **1 gallon**   Portion Size **6 oz.**

INGREDIENT	QUANTITY	COST A.P. $	COST Yield %	COST E.P. $	RECIPE COST
Onions, dice	3¾ lb.				$ 2.10
Leeks, chop	3				$ 4.35
Butter	2 oz.				$ .15
Flour	2 oz.				$ .03
Milk	1 qt.				$ .57
Salt	to taste				

TOTAL RECIPE COST $ **7.20**
Number of Portions **21**
Cost per Portion $ **.34**

CHAPTER 3

Labor Costs

The term ***direct labor*** refers to the work that is directly involved in the production of menu items. For example, the time that a cook spends cleaning lettuce, frosting a cake or shucking oysters would be direct labor used in producing those foods. The labor required to clean the parking lot, balance the cash register, inventory the wine cellar and so on is ***indirect labor***.

The cost of direct labor should be considered when costing menu items. If food is purchased fully processed, cleaned, portioned, and ready to heat and serve, labor costs will be low. The more in-house processing needed, the more labor costs will increase. At some point, the money saved by buying wholesale cuts of meat or making your own stocks may be outweighed by the costs of labor needed to produce those items. For example, if a prep cook works slowly and takes too long to clean a beef tenderloin, the chef may decide it would be less expensive to purchase the tenderloin already cleaned and portioned for service. The labor expense for that cook could then be eliminated.

Unit and Recipe Costing

> **?** Your manager wants to make all dinner rolls in-house, instead of buying ready-to-serve bread. He asks you if this is a good way to save money. What information do you need before you can answer him?

The expense for direct labor is added to food cost to arrive at a figure known as *prime cost*, which is the total cost of preparing a food item for sale. Prime cost is often used when setting menu prices, as described in Chapter 7, *Menu Pricing*.

Direct Labor + Food Cost = Prime Cost

Unit and Recipe Costing

Review Questions for Chapter 3

Define the following terms:

UNIT COST

COST PER PORTION

YIELD

A.P.

1. Why is it important to calculate the portion cost of a recipe in professional food service operations? Why is the cost of the recipe alone inadequate?

2. Explain how the cost of a small quantity of an ingredient can be determined from the known cost of a large or bulk quantity of that ingredient.

CHAPTER 4

YIELD TESTS

LEARNING OBJECTIVES:

- Understand how to calculate yield factors

- Understand how to use yield factors in purchasing

- Understand how to use yield factors in pricing

CHAPTER 4

Ingredient Yields

Computing the cost of recipe ingredients is a simple matter if foods are used the way they are received and there is no waste or trim. This is rarely the case, however. The amount of a food item *As-Purchased* (A.P.) and the amount of the *Edible Portion* (E.P.) of the same item may vary considerably, particularly with meat, fish, poultry and fresh produce. In this context, *yield* refers to the usable or edible quantity remaining after processing the as-purchased quantity of a food item. That is, yield refers to the amount of usable lettuce after the case of iceberg is cleaned, or the amount of meat that is available after trimming.

The *yield factor* or *yield percentage* is the ratio of the usable quantity to the purchased quantity. It is always less than 100% and may be calculated in dollars or quantity (weight/volume) amounts.

> **?** One of your cooks completes a yield test on a case of spinach and reports that spinach has a yield of 117%. What's wrong with his results?

Yield Tests

Because purchase specifications and fabrication techniques vary from operation to operation, there are no precise, standard yield amounts. Each kitchen should determine its own yield factors using its own cooks and its own cleaning and trimming standards. To be accurate, several tests should be conducted and the results averaged to arrive at a specific operation's yield factor for each item.

The method of calculating yield varies depending on whether the item's trim is all waste (for example, vegetable peelings) or whether the trim creates usable or saleable by-products (for example, meat and poultry). This chapter examines both types of yield tests and explains how this information is critical in purchasing control.

Raw Yield Tests Without By-Products

The simplest yield test procedure is for items that have no usable or saleable by-products. These foods include most produce as well as some fish and shellfish. Unless these foods are ready to serve as received from the purveyor, trimming is required and all trim is waste. For example, one pound of apples may yield only 13 ounces of flesh after peeling and coring. If a recipe requires one pound of peeled, cored apples, the chef must start with slightly more than one pound of A.P. apples. In order to determine accurate costs for such

CHAPTER 4

items, the trim loss must be taken into account. Even seedless grapes do not have a 100% yield factor. The weight of stems and bad grapes should be calculated and deducted from the A.P. weight to determine the correct price per pound of servable fruit.

Three steps are used for calculating yield when all trim is waste:

STEP 1: *Weigh* the trim produced from the specified A.P. quantity. This is known as *trim loss*.

STEP 2: *Subtract* the trim loss from the A.P. weight to arrive at the total yield weight.

STEP 3: *Divide* the yield weight by the A.P. weight to determine the yield factor (yield percentage).

Example 4.1

Two pounds of fresh garlic generates 4-1/2 ounces of trim loss. Therefore, the yield weight is 27-1/2 ounces and the yield factor is 86%:

$$(2 \text{ lb.} \times 16 \text{ oz.}) = 32 \text{ oz.} - 4.5 \text{ oz.} = 27.5 \text{ oz.}$$

$$\frac{27.5}{32} = .859 = 86\%$$

Yield Tests

Note that the *yield factor will always be some number less than 1* and the *yield percentage will always be less than 100%*.

$$A.P. - Trim\ Loss = Yield$$

$$\frac{Yield}{A.P.} = Yield\ \%$$

Practice calculating yield factors by completing Exercise 4.1. Remember, *subtract* trim loss from A.P. weight; then *divide* yield weight by A.P. weight to arrive at the yield percentage.

CHAPTER 4

Exercise 4.1

CALCULATING YIELD PERCENTAGES

Item	A.P. Weight	Trim Loss	Yield Weight	Yield %
Asparagus	14.5 lb.	5.5 lb.	9 lb	62%
Broccoli	27 lb. 432	10 lb. 10 oz. 170	26 2oz	61.%
Cabbage	18-1/4 lb. 292 oz	3 lb. 12 oz. 60oz	232oz	79%
Cucumbers	8 lb. 128oz	21 oz.	107oz	84%
Honeydew	6 2-lb. melons	20.5 oz	8.5 lb.	71%
Onions	25 lb.	47.25 lb.	22 lb. 4 oz.	89%
Leeks	6-2/3 lb. 4lb 64oz	3 lb. 4 oz	1 lb 14oz	55%
Tomatoes	18.75 lb. 300oz	19 oz.	281oz	94%

Because each operation has its own standards for cleaning and trimming raw products, yield factors should be personalized. Lists of common yield factors, such as those given in Chart 4.1, are available, however, as an indication of industry norms.

Yield Tests

Chart 4.1

COMMON PRODUCE YIELD FACTORS

Apples	75%	Grapefruit	45%	Pears	75%
Apricots	94%	Grapes	90%	Pea pods	90%
Artichokes	48%	Kiwi	80%	Peppers	82%
Avocados	75%	Leeks	50%	Pineapple	50%
Bananas	70%	Lemons	45%	Plums	95%
Berries	95%	Lettuce	75%	Potatoes	80%
Broccoli	70%	Limes	45%	Radishes	90%
Cabbage	80%	Melons	55%	Rhubarb	85%
Carrots	78%	Mushrooms	90%	Scallions	65%
Cauliflower	55%	Nectarines	86%	Spinach	60%
Celery	75%	Okra	82%	Squash	90%
Cherries	82%	Onions	90%	Tomatoes	90%
Corn, cob	28%	Oranges	60%	Zucchini	90%

According to Chart 4.1, 70% of the broccoli you purchase is usable, 30% is waste; 65% of all scallions are usable, 35% is waste. Depending on your employees' skills, cleaning standards, and purchasing specifications, these percentages may or may not be accurate for your operation.

CHAPTER 4

Applying Yield Factors

Now that you understand what yield factors are and how to calculate them, let's look at the two ways in which they are applied. First, yield factors are used for accurate ingredient pricing and recipe costing. Second, yield factors are necessary for accurate purchasing.

▶ To Determine Ingredient Price

When a food product is cleaned or trimmed, a portion of what you paid for is discarded. The cost of the remainder of that food should be adjusted upward to account for its increased value. This revised price is then the accurate one to use in recipe costing. In other words, the price should be recalculated to account for waste. The A.P. (as-purchased) unit cost must be converted to an E.P. (edible portion) unit cost. This is done by *dividing* A.P. cost by the yield percentage.

$$\frac{A.P.\ Cost\ (\$)}{Yield\ Percentage} = E.P.\ Cost\ (\$)$$

Yield Tests

Example 4.2

Carrots cost $6.50 per 25-lb. bag and have a yield factor of 78%; in other words, 22% of that 25-lb. bag is waste. So, the carrots have an E.P. unit cost of 33¢ per pound.

$$\frac{.26/\text{pound}}{.78} = .33/\text{pound}$$

When costing a recipe, 33¢ per pound is the accurate ingredient cost.

> **!** E.P. cost is always greater than A.P. cost.

Exercise 4.2 allows you to practice calculating the revised E.P. cost of food items that have no usable trim or by-products. Use the yield factors listed in Chart 4.1 to complete this exercise.

CHAPTER 4

Exercise 4.2

DETERMINING EDIBLE-PORTION COSTS

Item	A.P. Cost	Yield %	E.P. Cost
Asparagus	.70 lb.	74%	$.95
Avocado slices	1.29 lb.	75%	$1.72
Banana slices	.37 lb.	70%	$.53
Broccoli pieces	.89 lb.	70%	$1.27
Yams, diced	.08 lb.	80%	$.10

► To Determine Purchase Amount

Most recipes list ingredients in E.P. quantities. Therefore, the chef must consider waste or trim amounts when ordering these items. If the amounts listed in the recipe are ordered and then require trimming, the number of portions (or recipe yield) will be less than the desired amount.

Yield Tests

$$\frac{E.P.\ Quantity}{Yield\ Percentage} = A.P.\ Quantity$$

Example 4.3

A recipe requires 20 lb. of shredded cabbage. The yield factor for cabbage is 79%. Therefore, 20 lb. is 79% of the A.P. quantity. *Divide* the amount needed by the yield factor to determine the minimum A.P. quantity.

$$\frac{20\ lb.}{.79} = 25.3\ lb.$$

It will take 25-1/3 lb. of cabbage to provide the 20 lb. of shredded cabbage. (This figure should be increased to an even amount for purchasing because yield factors are, at best, only an estimate.) Note that the A.P. figure must always be greater than the E.P. figure in this formula.

CHAPTER 4

> **!** EP Quantity is always less than A.P. Quantity

Exercise 4.3 provides practice in calculating purchasing quantities for items having unusable trim. Remember that yield factors are listed in Chart 4.1.

Exercise 4.3
DETERMINING AS-PURCHASED QUANTITIES

Item	Recipe Amount	Yield %	A.P. Amount
Asparagus	3 lb.	74 %	4.05
Avocado slices	1-1/2 lb.	75%	2 lb
Banana slices	5.75 lb.	70%	8.2 lb
Broccoli pieces	2 lb.	70%	2.9 lb
Yams, diced	22-2/3 lb.	80%	28.3 lb

Raw Yield Tests With By-Products

With meats, poultry, and some fish, only a small amount of trim will be discarded as waste. Much of the trim can be used and therefore has a value to the operation. Calculating yields on food items with usable or salable trim is a bit more complex than calculating yields for food items whose trim is all waste.

If the food service operation purchases only precut, portion-controlled meats and seafood and uses them just as they are received, A.P. and E.P. costs are the same. But if the operation buys wholesale cuts of meat or whole fish and poultry and cuts them in-house, the chef should do a yield cost analysis to determine the actual costs.

The labor costs associated with trimming or fabricating food items can become a major expense for a food service operation. Although labor costs should be considered when making purchasing and production decisions, for now we are only concerned with actual *food* costs. The impact of labor costs was discussed in Chapter 3.

Example 4.4

Suppose your restaurant serves a boneless, skinless grilled chicken breast entrée. The restaurant buys whole fryers for $1.07 per pound. You must clean the chickens by removing the breast meat and skin, and separating the wings, legs, and thighs. You begin with a

CHAPTER 4

case of chickens weighing 35 lb. After fabrication you have 10-1/2 lb. of boneless, skinless breast meat. How do you calculate the cost per pound of this meat?

Calculations are simple if you discard all the trimmings, bones, and scraps. You divide the cost of the case (35 lb. × $1.07 per pound = $37.45) by the weight of the breast meat (10.5 lb.) to find the cost per pound ($3.57). But most restaurants that fabricate their own meats don't throw away the trimmings. Bones are used for stock, wings are served on the happy hour buffet, legs and thighs will be used for chicken salad or employee meals. Each of these secondary uses has some value, which must be included in food cost calculations.

Chart 4.2 is a typical form used for calculating yield. It contains the results from the above chicken breast fabrication. *(A blank Yield Test Form is included in the Appendix.)* The values for the by-products should be what it would currently cost to purchase them. These prices can be obtained from invoices or purveyors.

Some types of fat can be sold to a fat collection company. If your operation sells its fat, this portion of the trim would also have a value and would be included as a by-product. On the other hand, if your restaurant must pay a special waste removal company to discard large quantities of fat, the cost of fat removal should be calculated and included in your yield analysis.

A new term appears on the form: ***cutting loss***. This is not an item that is actually weighed. When you

Yield Tests

total the weights of the by-products and the tested item's yield, you find that the total is 2-1/2 lb. less than the A.P. weight (35 lb.). That 2-1/2 lb. can be accounted for by particles of meat and fat sticking to the cutting board, blood and moisture loss and other factors. Cutting loss is unavoidable and has no value. It is included on the chart as a "double-check" for other calculations.

Chart 4.2

YIELD TESTING FORM

Food Item **Chicken Breast** Date **11-26**

A.P. Price★ **$1.07** A.P. Weight★ **35 lb.**

Total A.P. Cost★ **$37.45**

Total Item Yield **10.5 lb.** Total Net Cost ❶ **$22.50**

Net Cost per Pound ❷ **$2.14** Percent of Increase ❸ **200%**

By-products, Trim and Waste:

Item	Weight	$ Value	Total $ Value
Wings	4.5 lb.	.60	2.70
Legs & Thighs	12.5 lb.	.90	11.25
Bones	5 lb.	.20	1.00
Cutting Loss	2.5 lb.	—0—	—0—

Total Weight **24.5 lb.** Total Value $ **14.95**

CHAPTER 4

After calculating the total value of all by-products, subtract that amount from the original cost of the chicken as-purchased. This figure reflects the *net cost* of the 10.5 pounds of boneless, skinless breast meat.

$$\$37.45 - 14.95 = \$22.50$$

To find the cost per pound for the breast meat, divide the net cost by the tested item's yield weight. This is the figure you will use for costing chicken breast recipes.

$$\$22.50 \div 10.5 = \$2.14 \text{ per pound}$$

Percentage of increase is determined by dividing the net cost per pound by the A.P. price per pound. Note that this figure is almost always greater than 100%.

$$\$2.14 \div \$1.07 = 200\%$$

The percentage of increase figure is used the next time the operation purchases chicken but finds that the price has changed. Instead of completing another yield test, the new cost per pound for chicken breast meat can be determined by multiplying that *new* price by the percentage of increase. So, if the cost of whole chickens rises to $1.85 per pound, breast meat will cost $3.70 per pound:

$$\$1.85 \times 200\% = \$3.70$$

> **?** What happens to the cost of breast meat if the cost of whole chickens falls to 98¢ per pound?

Cooked Yield Tests

So far, this chapter has addressed yields of uncooked food items. Most foods, however, suffer additional loss in the cooking process. Loss also occurs when trimming fat or bones, and when slicing and portioning foods after cooking. To accurately determine portion cost and purchasing quantities, these losses must be considered, especially if the food portion is determined by weight after cooking (for example, sliced roast beef for sandwiches). Once again, several tests should be conducted to obtain reliable yield factors for each operation. *As-Served* (A.S.) refers to the actual weight of one portion as it is served to the customer. If a food is served raw, such as fruit, the as-served weight will be the same as the edible portion weight. If the food is cooked, however, these weights will be different.

Chart 4.3 is a Cooked Yield Testing Form used for determining cooked cost per pound and percent of shrinkage. Remember that cooked yield tests are only

CHAPTER 4

necessary where A.S. portions are based on weights *after cooking*. If, for example, sautéed shrimp is cooked to order based on a 6-oz. uncooked portion, a cooked yield test is not necessary. If, however, large quantities of shrimp are cooked first, then portioned for service, a cooked yield test must be completed to determine the actual cost per portion.

Chart 4.3

COOKED YIELD TEST FORM

Item __Turkey Breast__ Date __4-18__

Net weight __18 lb.__ Net Cost per Pound __$3.30__ Net Cost __$59.40__

Portion Size __6 oz.__ Portions Served __40__
Total Weight As-Served __15 lb.__
Cooked Cost per Pound __$3.96__
Shrinkage __3 lb.__ Percent of Shrinkage __17%__
Total Percent of Increase _____

Yield Tests

Chart 4.3 is filled in with the results of a cooked yield test on a roasted boneless turkey breast. Assume that a raw yield test has already been conducted on the turkey breast, providing the net cost and net weight figures. The weight as served is calculated by keeping track of the number of portions actually served from that turkey and multiplying that number by the portion size. While it is tempting to simply weigh the turkey after cooking and trimming, this fails to account for loss caused by employee snacking, wasted scraps, crumbs left on the cutting board or slicer, or spilled juices. It is more accurate to record the weight that is actually served.

The remaining lines on the form are completed with your calculations, just like on the raw yield test form. The cooked cost per pound is the net cost *divided* by the A.S. weight ($59.40 ÷ 15 lb. = $3.96).

Shrinkage refers to the loss from pre-cooked weight to as-served weight (18 lb. - 15 lb. = 3 lb.). The ***percent of shrinkage*** figure is the ratio of shrinkage to raw weight (3 lb. ÷ 18 lb. = 17%). It is used to assist the chef in determining future purchasing needs. For example, if the chef knows that roast beef has a 12% shrinkage factor, she knows that the pre-cooked weight must be 12% more than the weight needed after cooking.

Total percentage of increase shows the increase in cost per pound from purchasing to service. It is calculated as a follow-up to a raw yield test by dividing the cooked cost per pound by the A.P. cost per pound.

CHAPTER 4

> **?** If whole raw turkey costs $1.89 per pound, what is the percent of cost increase for cooked, sliced turkey breast? *(See Chart 4.3.)*

Using Yield Tests for Purchasing

As discussed above, yield factors are useful in determining how much of a raw item to buy when you know how much of the cooked item you need. The calculations for items with by-products and for cooked items are the same as shown in the produce examples above.

STEP 1: Calculate the yield percentage.

E.P. ÷ A.P. = Yield Percentage

STEP 2: Divide the amount needed (E.P.) by the yield percentage to determine the amount to purchase.

E.P. Quantity ÷ Yield % = A.P. Quantity

Yield Tests

Example 4.5

Refer to the chicken breast example in Chart 4.2 and assume that you need 43 lb. of raw chicken breast meat for a dinner party. The raw yield factor is 30% (10.5 ÷ 35 = .30). So, you must purchase at least 143-1/3 lb. of whole chickens to have the 43 lb. of breast meat needed (43 ÷ .30 = 143.33).

Exercise 4.4

Return to the sliced turkey example above and assume that you need to make 75 turkey sandwiches, each containing 4 oz. of cooked, sliced turkey. If you know that a 24-lb. turkey produces 18 lb. of raw turkey breast, how many pounds of whole, raw turkeys will you need to buy to have enough breast meat for the sandwiches?

> ! A.P. = As-Purchased
> E.P. = Edible-Portion
> A.S. = As-Served

CHAPTER 4

Making Purchasing Decisions

Deciding to purchase food items that are ready-to-serve, or deciding to do the preparation in-house, involves more than a just a cost analysis. Other factors may be significant in determining the best approach for your operation. Consider the following questions:

Employee Skills Do your employees have the skills necessary to butcher meats properly? To fabricate fish? To clean and prepare vegetables consistently? How much will hiring skilled labor cost?

Menu Do you have a use for the bones, meat, and trimmings that result from fabricating large cuts of meat into individual portions? Can your menu change depending on the seasonal availability of fresh produce or shellfish?

Storage Do you have ample refrigerator and freezer space so that you can be flexible in purchasing highly perishable items, such as produce, meat, poultry, and seafood?

Customers Do your customers expect fresh produce and custom cuts of meat? Do they prefer desserts made "from scratch"? Are they willing to pay for additional labor costs?

Yield Tests

Review Questions for Chapter 4

Define the following terms:

AS SERVED

YIELD FACTOR

CUTTING LOSS

1. What is the purpose of the Percentage of Increase figure on the raw yield test form? How is this figure used by the chef?

2. List several factors that might cause one restaurant's yield factor for lettuce to be higher than another restaurant's. Explain why standardized yield factor lists are unreliable.

CHAPTER 5

INVENTORY AND FOOD COST PERCENTAGES

LEARNING OBJECTIVES:

- Understand how and why to conduct an inventory

- Understand and be able to calculate cost of goods sold

- Understand and be able to calculate food cost percentages

CHAPTER 5

Inventory

An inventory is an itemized list of the goods or merchandise owned by a business at a specific point in time. An accurate inventory is necessary in order to calculate financial statements, and can be used to pinpoint problems and predict future needs. A restaurant's inventory may include deep fryers, bar glasses, table linens, and stock pots, but this discussion focuses only on the food and beverage items used to produce items for sale. Foods and beverages can be inventoried on an on-going basis and through a periodic physical inspection and counting.

> **!** INVENTORY = Listing, counting and valuing all foodstuffs held by an operation at a given point in time.

A *perpetual inventory* is one maintained on an on-going basis using a record book, shelf cards (see Chart 5.1), or computer scanning. It shows the balance on hand for each item at any time. This is sometimes referred to as a running inventory. As an item is issued or used it is subtracted from the book or card; when an item is received it is added to the book or card. This can be done manually or by scanning the bar code printed on each item. Maintaining such a system requires either a sophisticated computer network or a great deal of labor; therefore, it is generally only used in large operations. However, even small facilities may find it beneficial to maintain a perpetual inventory for high-value items (such as meat), high-volume items (such as coffee), and regulated items (such as alcoholic beverages). In these cases, the cost is often outweighed by the benefit of having inventory information constantly available.

Chart 5.1

SAMPLE PERPETUAL INVENTORY CARD

Item _____   Bin # _____
Size _____   Par _____
Vendor _____   Reorder Point _____
_____   Reorder Amt. _____

Date	Invoice #	Received	Issued	Balance On-hand

CHAPTER 5

A *physical inventory* requires listing and counting all foods in the kitchen, storerooms and refrigerators. The quantities are then *extended;* that is, multiplied by the unit cost. The extended prices are then added to calculate the total inventory value. A physical inventory should be taken whenever management needs an accurate food cost figure. This can be at the end of each week, month, quarter or other accounting period.

To conduct a physical inventory, a standardized form such as the one in Chart 5.2 is created. Items are listed on the form in the order in which they are stored on shelves. Two people work together to take the actual inventory. One person counts and calls out items and quantities; the other records this information on the form.

Whether prepared foods and open containers are included in a physical inventory depends on the food service operation. Often small quantities of prepared foods or open containers are not inventoried on the theory that a certain amount of such items is always on hand, the value of which is fairly consistent from one accounting period to the next.

Inventory and Food Cost Percentages

Chart 5.2

SAMPLE INVENTORY FORM

STOREROOM INVENTORY				
June 30, 19xx			By	*DM*
ITEM	PACK	QUANTITY	PRICE	TOTAL
Tomato paste	#10	18	3.85	69.30
Tomato sauce	#10	6	3.45	20.70
Spaghetti sauce	#10	5	5.87	29.35
Clam juice	46 oz.	12	4.37	52.44
Artichoke hearts	#10	3	9.87	26.91
Tomato juice	46 oz.	8	2.39	19.12
Apple juice	46 oz.	5	1.89	9.45
				229.97

Shrinkage

In an operation that maintains a perpetual inventory, a periodic physical inventory will also serve to highlight any shrinkage that may occur. ***Shrinkage*** is the difference between what should be on hand based on purchases and usage and what is actually on hand by physical count. Shrinkage losses may be caused by outright theft, collusion with vendors, poor record-keeping, or spoilage.

CHAPTER 5

> SHRINKAGE = The difference between the amount of assets that should be on-hand and the amount that is actually on-hand.

Exercise 5.1

On the first day of March there were three bottles of pinot noir wine in the storeroom. One new case (12 bottles) was received during each of the following three weeks. One bottle of the wine was sold every night (31 bottles) during March. What should the perpetual inventory bin card indicate?

The physical inventory at the end of March showed 5 bottles on the shelf. What does this indicate?

Controlling Shrinkage Shrinkage, caused by either mishandling of products or outright theft, can have a devastating effect on an operation's bottom line. It is, therefore, important to establish and follow security measures aimed at reducing these losses. For example, deliveries should be checked in and properly stored as soon as possible. A case of beef tenderloins left unattended on the loading dock not only suffers from the lack of refrigeration. It may offer more temptation than some employees can resist.

Inventory and Food Cost Percentages

Shrinkage can also be reduced or eliminated by properly screening potential employees, training employees in procedures for storing, requesting and issuing foodstuffs, and demonstrating zero-tolerance for snacking, pilferage, or theft. To control loses it may be necessary to:

- maintain locked or guarded storage areas,
- randomly search dumpsters and trash bins,
- examine all employee packages leaving the premises,
- conduct unscheduled or irregular inventories,
- centralize cashiering functions, especially when offering promotional coupons, and
- watch employee exits and entrances at all times.

> **?** What might motivate or cause an otherwise honest employee to steal?

Cost of Goods Sold

Perhaps no other cost is emphasized as much by food service managers as *food cost*. Food cost refers to the cost of all foods and beverages used in the

83

CHAPTER 5

fabrication of menu items. This figure is also known as the ***cost of goods sold*** or raw food cost. The term "raw" is not meant literally, of course, as food cost includes pre-cooked and packaged foods as well as uncooked foods. The "goods" sold by a food service operation are, of course, the foods and beverages used in producing menu items.

> **!** FOOD COST = The cost of all the foods and beverages used in producing menu items.

Food costs are calculated in two ways: (1) as the total cost of all foods used during a given time period, and (2) as the cost of one portion or menu item. Total cost is used as a general guideline for budgeting and menu planning. The portion or item cost is used to calculate menu prices and helps the chef stay within cost limitations.

To properly calculate the cost of goods sold (CGS) you must conduct an inventory at the beginning and end of the desired period and maintain records of all purchases during the period. Generally, supplies, tools, and non-food items are not included in CGS calculations, although this varies depending on the way such non-food items are used. The time period covered

Inventory and Food Cost Percentages

by this calculation could be any duration—day, week, month, quarter, or year—depending on the information needed by management.

After the total value of the inventory for both the beginning and the end of the period have been established, cost of goods (food) sold* is calculated as follows:

	Value of Food Inventory at Beginning of Period
PLUS	*Value of Food Purchased During Period*
MINUS	*Value of Inventory at End of Period*
	Cost of Goods (Food) Sold

Example 5.1

The total value of food in inventory on December 1 is $7,600. The restaurant purchases $2,300 worth of food during December and the inventory on January 1 is worth $5,600. The cost of goods sold

---

* All of the foodstuffs taken from inventory may not be actually sold. They may be used for employees meals, transferred to the bar, sold in a retail market, and so on. A more accurate term, therefore, might be cost of goods *consumed*. Accounting for such uses is beyond the scope of this text.

CHAPTER 5

during the month of December is $4,300 (in other words, the food produced during the month of December cost $4,300):

$$(\$7{,}600 + \$2{,}300) - \$5{,}600 = \$4{,}300$$

When management knows the cost of goods sold, it can compare this figure with the dollar value of sales for the same period. This comparison gives management a good idea of how the business did during that time period. These figures can also be compared with the cost of goods sold and sales for prior periods, such as the same month in the previous year.

Exercise 5.2

The value of foodstuffs on hand on January 1 is $11,400. Inventory on the following December 31 is $13,300. During the year your restaurant purchased foodstuffs valued at $120,200. What was the cost of goods sold for that year?

Food Cost Percentages

The *food cost percentage* is the ratio of costs to sales. It shows what each dollar of sales costs. For example, a 35% food cost means that 35¢ of each dollar received went to pay for the foods the operation used.

86

Inventory and Food Cost Percentages

Food cost percentage is determined by *dividing* food cost by sales.

$$\frac{\textit{Cost of Food Sold}}{\textit{Food Sales}} = \textit{Food Cost Percentage}$$

Example 5.2

Refer back to Example 5.1 and assume that food sales for December totaled $10,750. The food cost percentage for the month would be 40%:

$$\frac{4{,}300}{10{,}750} = .40 = 40\%$$

> **!** FOOD COST PERCENTAGE = The ratio of the cost of food served to the food sales income received during a given period of time.

87

CHAPTER 5

Exercise 5.3

Refer back to your answer in Exercise 5.2. Assuming that sales for that year were $340,000, what was the annual food cost percentage?

By itself a single food cost percentage is meaningless. To be useful, it should be compared with the food cost percentages for other months of the same year or the same month in previous years. It is also helpful to know the food cost percentages for other similarly situated food service operations. For example, convenience or fast food type restaurants tend to run a much higher food cost percentage than fine dining establishments. In fast food operations 35 to 45% might be considered normal or acceptable; other restaurants consider a food cost of less than 30% necessary. There is no perfect food cost percentage—each food service operation must determine its own *ideal* percentage based on a variety of factors. Considerations might include: the availability and cost of labor, the amount of competition and the prices charged by that competition, and overhead expenses.

Food cost percentages can also be calculated on individual menu items. If the food items in a sliced turkey sandwich cost $2.80 and the sandwich sells for $5.25, the food cost percentage is 53% ($2.80 ÷ $5.25). Most operations would find this percentage unacceptably high. To reduce the percentage and increase gross profits, the operation must either increase the price or decrease the cost of ingredients.

Inventory and Food Cost Percentages

Changing menu price, raw food cost, or portion size will change the food cost percentage. If the food cost percentage is too high, gross profit will be insufficient to cover operating expenses. If the percentage is too low, customers may feel they are not getting their money's worth and take their business elsewhere. Setting an appropriate percentage requires sound judgment, knowledge of the competition, and accurate cost information. Periodic evaluations are necessary to ensure that the desired objective is actually maintained.

Food cost percentages are also used to determine the selling, or menu, price for foods. This subject is covered in Chapter 7, *Menu Pricing*.

Inventory and Food Cost Percentages

Review Questions for Chapter 5

1. At the beginning of the month there were three cans of escargot on the shelf. Each can cost $12.57. During the month two cases containing 24 cans each were purchased and 43 cans were used by the kitchen. What should the end of month physical inventory show?

2. How often should a restaurant conduct a physical inventory of foods on hand? Why?

3. A fast food restaurant wraps every sandwich it sells in a sheet of wax paper. This paper costs $21.00 per 1000. Should the inventory of wax paper be included in the cost of goods sold? Why?

4. Calculate the food cost percentage for an operation with annual sales of $540,000. The physical inventory at the start of the year listed foods on hand valued at $23,000; the year end inventory listed foods valued at $17,000. Purchases during the year totaled $235,000.

CHAPTER 6

CONTROLLING FOOD COSTS

LEARNING OBJECTIVES:

- Understand the internal and external factors affecting food costs

- Understand the role of training in controlling food costs

CHAPTER 6

Many things affect food costs in any given operation. Most factors, however, can be controlled by the chef or manager. These controls do not require mathematical calculations or formulas, just basic management skills and a good dose of common sense. The following factors all have an impact on the operation's bottom line:

- menu

- purchasing/ordering

- receiving

- storing

- issuing

- kitchen procedures

    - establishing standard portions

    - waste

- sales and service

Chefs tend to focus their control efforts in the area of kitchen precedures. While this may seem logical, it is not adequate. A good chef will be involved in all aspects of the operation to help prevent problems from arising or correct those that may occur.

Controlling Food Costs

Menu A profitable menu is based upon many variables, including customer desires, physical space and equipment, ingredient availability, cost of goods sold, employee skills, and competition. All management personnel, including the chef, should be consulted when planning the menu. Menu changes, while possibly desirable, must be executed with as much care as the original design. A menu that changes too dramatically or too frequently may result in unusable leftovers or an inventory of unneeded raw ingredients.

Purchasing and Ordering Purchasing techniques have a direct impact on cost controls. On the one hand, a parstock of supplies must be adequate for efficient operations; on the other hand, too much inventory wastes space and money, and it may spoil. Before any items are ordered, purchasing specifications should be established and communicated to potential purveyors. Specifications (specs) should precisely describe the item, including grade, quality, packaging, and unit size. *(See Chart 6.1)* This information can be used to obtain price quotes from several purveyors. Update these quotes periodically to ensure that you are getting the best value for your money.

! PARSTOCK (PAR) = The amount of stock necessary to cover operating needs between deliveries.

CHAPTER 6

Chart 6.1

SAMPLE PURCHASING SPEC FORM

Pima Grill	
MEAT PURCHASING SPECIFICATIONS	
Product:	
Menu Item:	
Grade/Quality:	**NAMP/IMPS #:**
Packaging:	
Pricing Unit:	
Delivery Conditions:	
Comments:	

Controlling Food Costs

Receiving Whether goods are received by a full-time clerk, as they are in a large hotel, or by anyone who happens to be available at the time of delivery, as is more often the case, certain standards should be observed. The person signing for the merchandise should first confirm that the items were actually ordered. Second, confirm that the items listed on the invoice are the ones being delivered and that the price and quantity listed is accurate. Third, the items being delivered must be checked for quality, freshness, and weight. Established purchase specifications should be available in case questions arise.

Storing Proper storage of foodstuffs is crucial in order to prevent spoilage, pilferage and waste. Stock must be rotated so that the older items are used first. Such a system for rotating stock is referred to as FIFO— First In, First Out. Dry storage areas should be well ventilated and well lit to prevent pest infestations and mold. Freezers and refrigerators should be easily accessible, operating properly, and kept clean and organized.

> ! FIFO = First In, First Out
>
> LIFO = Last In, First Out

CHAPTER 6

Issuing It may be necessary, particularly in larger operations, to limit storeroom access to specific personnel. Foodstuffs are then requisitioned by the kitchen and issued (or delivered) to the cooks as needed. Controlling issuances eliminates waste caused by multiple opened containers and ensures proper stock rotation. The ongoing inventory records and parstock sheets also assist with the ordering process.

Kitchen Procedures — *Establishing Standard Portions* Standardizing portions is essential to controlling food costs. Unless portion quantity is uniform, it will be impossible to compute portion costs accurately. Portion discrepancies can also confuse or mislead customers.

Actual portion sizes depend on the type of food service operation, its menu, prices, and customers' desires. Some items are generally purchased pre-portioned for convenience (for example, steaks are sold in uniform cuts, baking potatoes are available in uniform sizes, butter comes in pre-portioned pats, and bread comes sliced for service). Other items must be portioned by the establishment prior to service. Special equipment makes consistent portioning easy. There are machines to slice meats, cutting guides for cakes and pies, and portion scales for weighing quantities. Standardized portion scoops and ladles are indispensable for serving vegetables, soups, stews, salads and similar foods. (See the Appendix for detailed information.)

Controlling Food Costs

Once acceptable portion sizes are established, employees must be trained to portion foods properly and consistently. If each employee of a sandwich shop prepared sandwiches the way he or she would like to eat them, customers would probably never receive the same sandwich twice. Customers may become confused and decide not to risk a repeat visit. Obviously, carelessness in portioning can also drastically affect food cost.

Kitchen Procedures — *Waste* The chef must also control waste from overproduction or failure to use leftovers. The less waste generated in food preparation, the lower the overall food cost will be. If the menu is designed properly, the chef can use leftovers and trim items from product fabrication. With an adequate sales history, the chef can accurately estimate the quantity of food to prepare for each week, day, or meal. Physically counting and charting production will create an accurate sales history. Chart 6.2 is designed to create a sales history and to show where pilferage problems may be occurring.

When completing a daily report, the **amount to produce** column lists the amount the chef or manager has determined should be produced, while the **amount produced** column shows the amount *actually* produced. Discrepancies may be used to point out problems in inventory control, communication or time management.

For each menu item the **number sold** should be taken from guest checks or cash register records; it should not be the number prepared by the kitchen. This

CHAPTER 6

will allow the report to serve as a double-check on sales procedures and, when compared with a physical inventory of **leftovers**, will highlight areas of potential pilferage, waste, or avoidable loss.

Chart 6.2

DAILY PRODUCTION REPORT

Department __Bakery__   Date/Shift __3-16 / Lunch__

Menu Item	Amount Ordered	Amount Produced	Number Sold	% of Sales	Leftovers
Cheesecake	24	24	24	37	0
Mousse Cake	24	24	14	22	8
Apple Tart	6	6	6	9	0
Flan	12	14	13	20	1
Fruit Plate	12	8	8	12	0
			TOTAL	65	100%

Controlling Food Costs

Percent of sales is calculated by dividing the number of each menu item sold by the total number of items sold in that category. For example, in Chart 6.2, the 14 servings of mousse cake were 22% of total dessert sales (14 ÷ 65 = .22). A history of this information makes it possible to predict what percent of customers will order each dessert on any given day.

> **?** What might have caused the discrepancy between the number of servings of mousse cake that should have been left over and the number that actually was left over?

Sales and Service An improperly trained sales staff can undo even the most rigorous of food cost controls. Front of the house personnel are, after all, ultimately responsible for the *sales* portion of the food cost equation. Prices charged must be accurate and complete. Poor service can lead to the need to "comp," that is, serve for free, an excessive amount of food. Dropped or spilled food does not generate revenue.

Once again, proper training is critical. The dining room manager and the chef should work together to educate servers about menu items and sales techniques.

Controlling Food Costs

Review Questions for Chapter 6

The seven factors discussed in this chapter are summarized below. Add your own ideas to the potential problems listed, then list and explain possible solutions.

Menu

GOAL: To provide the type and variety of food desired by customers at a price they are willing to pay.

PROBLEMS: Menu prices too high or too low; too many items on the menu.

SOLUTIONS:

Purchasing and Ordering

GOAL: To obtain the best quality products, following established specifications, at the lowest possible prices.

PROBLEMS: No specifications; lack of suppliers; seasonal price changes; prices not based on actual quality or yield.

SOLUTIONS:

CHAPTER 6

Receiving

GOAL: To obtain the quality and amount ordered at the price quoted.

PROBLEMS: Untrained personnel; lack of communication between chef and receiving clerks; no physical check of goods; employee theft.

SOLUTIONS:

Storage

GOAL: To have sufficient stock available for production, without loss from theft or spoilage.

PROBLEMS: Improper storage conditions; employee pilferage.

SOLUTIONS:

Controlling Food Costs

Issuing

GOAL: To account for food issued to authorized personnel.

PROBLEMS: No record keeping; excess amounts issued.

SOLUTIONS:

Kitchen Procedures

GOAL: To plan for and prepare the amount of food needed, while avoiding over-production or waste.

PROBLEMS: Lack of sales history; variations in quality and portion size; untrained personnel; pilferage; failure to use leftovers; lost sales due to unavailable product.

SOLUTIONS:

CHAPTER 6

Sales

GOAL: To receive payment for all food prepared.

PROBLEMS: Inaccurate ordering; incorrect charges to customer; pilferage; mishandled food.

SOLUTIONS:

CHAPTER 7

MENU PRICING

LEARNING OBJECTIVES:

- Understand how to calculate menu prices using a variety of methods

- Understand the psychological impact of prices

Chapter 7

After determining the cost of food items, you can more intelligently and accurately calculate menu prices. A few of the many methods for setting menu prices are explained below. Some techniques are highly structured and closely related to food costs; others are unstructured and unrelated to actual food costs. As a practical matter, no one method is right for every operation and a combination of methods may provide the best pricing strategy.

Cost-Based Menu Pricing

This section examines four cost-based pricing methods: food cost percentage, factor pricing, prime cost pricing, and perceived value pricing. Each of these methods uses the *cost* of the food item as the starting point.

Food Cost Percentage For this technique, you must first determine the food cost percentage desired for the particular operation. Calculate each item's raw food cost, then *divide by* the desired food cost percentage to determine the selling price.

$$\frac{\textit{Cost per Portion}}{\textit{Food Cost \%}} = \textit{Selling Price}$$

Menu Pricing

Example 7.1

If the cost for one sandwich is $1.70 and the desired food cost percentage is 23%, the selling price must be at least $7.39

$1.70 ÷ $.23 = 7.391

In this situation, 23% of the sales price for each sandwich will go to cover the cost of ingredients. The higher the food cost percentage, the lower the portion of the sales price that is available for fixed expenses, overhead or profit. Determining the appropriate food cost percentage for your facility is critical to the successful use of this method.

Exercise 7.1

Your manager has determined that food costs should be no more than 28% of sales. What should you charge for a bowl of onion soup if the recipe cost is $1.12 per portion ?

Factor Pricing A variation on food cost percentage pricing is factor pricing, in which a multiplier is used to calculate menu prices. First, take the desired food cost percentage and divide it into 100 to arrive at a cost factor. The cost of each menu item is then multiplied by this cost factor to arrive at the menu price. Using the figures in Example 7.1, the factor is 4.35 (100

Chapter 7

÷ 23 = 4.35) and the menu price for a $1.70 item is $7.39 ($1.70 × 4.35 = $7.39).

Exercise 7.2

What factor is used to set menu prices with a 28% food cost?

Both food cost percentage and factor pricing are fast and easy to use. These methods are sometimes unreliable, however, because they assume that other costs associated with preparing food stay the same for each menu item. These systems wrongly assume that the costs for labor, energy, and overhead are the same for a rack of lamb entrée and a seafood salad. These methods may be fine-tuned somewhat by adjusting the desired food cost percentage according to the type of food or menu category. For example, appetizers may be assigned a lower food cost percentage than desserts or side dishes.

Menu Pricing

> **?** Why might one category of foods, such as desserts, be priced with a higher food cost percentage than other menu items?

Prime Cost Pricing The food service industry uses a figure known as *prime cost* to refer to the total of raw food cost plus direct labor. Direct labor is the labor actually required for an item's preparation. If, for example, food cost is $2.10 and direct labor is $1.50, then prime cost is $3.60. As with the food cost percentage example above, the prime cost percentage can be divided into the prime cost amount to determine menu price.

> **!** DIRECT LABOR = The labor required for the actual production of menu items.

Chapter 7

Example 7.2

The raw food cost for a rack of lamb is $7.50 and it takes a cook a total of 9 minutes to clean and trim it for service. If that cook is paid $8.50 per hour, the direct labor cost is $1.27 ($8.50 per hour = $.14 per minute) ($.14 × 9 minutes = $1.27).

The prime cost of the rack of lamb is $8.77 ($7.50 + $1.27). A selling price is then determined by dividing prime cost by the desired prime cost percentage. If the desired prime cost percentage is 48%, the rack of lamb should be priced at $18.27 ($8.77 ÷ .48 = $18.27).

> **!**   PRIME COST   = Food Cost + Direct Labor
> PRIME COST % = Food Cost % + Direct Labor %

Perceived Value Pricing This is a rather backwards way of setting prices based on customer perceptions of appropriate prices. You first determine what the "market price" is for the same or similar item. Then calculate what *you* can serve for that price.

This technique requires a thorough knowledge of the market and complete, accurate historical information on labor and overhead costs. With the necessary

Menu Pricing

information and experience, you can determine how much to spend on raw product for each menu item and still make the desired profit. This method is most useful for standard food items or in markets that are highly competitive.

Example 7.3

Assume that the market price for a complete chicken dinner is $9.95. Assume further that the desired profit is 7%, overhead expenses are 27% of sales and labor is 23% of sales. You can then spend 43% of $9.95 ($4.28) on the food cost for that chicken dinner.

7 + 27 + 23 = 57%

100 − 57 = 43% × $9.95 = $4.28

Exercise 7.3

Your competitor serves a complete shrimp dinner for $12.99. You would like to meet this offer, but you're not sure you can afford to. You know that your overhead accounts for 38% of sales, labor is 34% of sales, and your desired profit margin is 12%. How much can you spend on raw food cost for a $12.99 shrimp dinner?

Chapter 7

Profit-Based Menu Pricing

Unlike cost-based pricing methods, profit-based methods use *desired profit* as the starting point for calculations. The two methods discussed here are gross profit pricing and budget-based pricing.

Gross Profit Pricing Gross profit pricing is based on the specific amount of profit that is historically produced by each customer. So, there must be a sales history for the facility before this method becomes useful. The average gross profit per customer is first determined by dividing total gross profit by the total number of customers during the given period. If, for example, total gross profit for the first year of operation was $120,000 and a total of 40,000 covers were served during the year, the average gross profit per customer was $3.00. So, $3.00 is added to the food cost for each item to determine the menu price.

$$\frac{120,000}{40,000} = 3.00 + \text{food cost} = \text{menu price}$$

This pricing method is probably most useful for *prix fix* or buffet operations where food cost is also averaged.

Example 7.4

During the past 12 months a total of 28,500 customers were served and gross profit was $80,000.

Menu Pricing

The average gross profit per customer is $2.80 ($80,000 ÷ 28,500 = $2.80). If the average per person food cost for a salad bar is $2.10, the menu price for that salad should be $4.90 ($2.10 + $2.80 = $4.90).

Exercise 7.4

The average per person food cost for your Sunday buffet is $6.45. During the past year a total of 5200 people purchased this buffet and your gross profit was $31,500. How much should you charge per person if you wish to maintain that profit margin?

Budget-Based Pricing Budget-based pricing builds a target profit percentage into the selling price of each menu item. As with gross profit pricing, there must be a history of expenses to use this method. The desired profit percentage and the historical percentage for overhead (total labor, fixed expenses, and so on) are subtracted from 100%. The percentage remaining is the amount available for food costs. A factor is determined by dividing this food cost percentage into 100. The actual food cost for each menu item is then multiplied by the factor to obtain the selling price.

Chapter 7

Example 7.5

If the desired profit is 12% of sales and overhead expenses historically total 60% of sales, then the food cost percentage must be no more than 28%.

$$100 - (60 + 12) = 28$$

Dividing 28 into 100 provides a pricing factor of 3.57. If the raw food cost for a Chicken Caesar Salad is $2.59 then the selling price is $9.25.

$$\$2.59 \times 3.57 = \$9.25$$

Exercise 7.5

Your bakery is located in a shopping mall with extremely high rents. Consequently, your total overhead averages 75% of sales. Your food cost for one dozen cranberry muffins is $1.04. If you want to make 14% profit, how much should you charge for one dozen muffins?

Non-Cost-Based Pricing

Non-cost-based pricing techniques rely on non-monetary factors in setting menu prices. Traditional prices in your area for the same or similar items may affect your ability to set prices. This is particularly true

Menu Pricing

for special or loss-leader items such as a 79¢ jumbo soda or a $1.99 breakfast special.

While the competition's prices are an important consideration, you should not simply copy them. Your competition's costs are not your costs no matter how similar the final food product appears. If you must charge higher prices than your competition, seek out some way to *differentiate* your product or your service.

Virtually all food service operations are seeking to charge the highest price possible without a loss of sales. The customer's perception of value is therefore critical. The menu can be used as a tool for educating customers. Descriptive language or an explanation of unique or special dishes can be included. Customers who understand the value of service, atmosphere, out-of-season foods, and specialty products will tolerate higher prices for those items.

Psychological Impact of Pricing

Regardless of how you arrive at a menu price, you may wish to round the figure up or down for psychological impact. It is human nature to perceive some prices as higher or lower than they actually are. For example, using a 9 or 5 as the **last digit** in a price creates the impression of a discount. Prices ending in a 9 or 5 appeal to price-conscious customers. On the other hand, prices ending in a 00 or 0 are perceived as more

Chapter 7

expensive and of higher quality and so are often used on fine-dining menus.

The *number of digits* in a price is also important. $9.95 seems much less expensive than $10.25. Likewise, the *first numeral* in a price affects perception. When changing menu prices, an increase from $5.95 to $6.45 seems greater than an increase from $6.25 to $6.75, although both are 50¢ increases.

Patrons do not like to see a large *spread in menu prices*. Erratic pricing may cause confusion or make customers think something is wrong with one or more of the items. For example, a lobster dinner for $18.50 may be a reasonable price, but it may be inappropriate on a menu where all other items are less than $6.00. In general, *the highest price should not be more than double the lowest price within the same food category.* If the least expensive appetizer is $4.00, then the most expensive appetizer should not be more than $8.00.

Exercise 7.6

Your accountant accurately calculated all your menu prices based on information you provided. Her prices show figures such as $6.12, $9.00, $8.57 and $1.19. Use your knowledge of pricing psychology to adjust these prices to more appropriate figures and explain your reasoning.

Menu Pricing

Review Questions for Chapter 7

1. In order to maintain an average food cost of 35%, what multiplier should you use to calculate menu prices?

2. Calculate the menu price for a chicken salad with a raw food cost of $1.97, where the desired food cost is 23%. How might this price appear on an upscale restaurant's menu?

3. During June your restaurant will offer a complete shrimp dinner for a special price of $8.99. Explain how you will determine the appropriate cost for the raw ingredients for this meal. What other information do you need?

Chapter 7

4. The five appetizers on your lunch menu are priced from $6.50 to $7.95 each. Your manager wants to add a new appetizer priced at $2.95. Explain why you disagree with this idea.

5. Why might it be important to know the labor cost directly associated with preparing a given food item? Explain how the prime cost percentage differs from the food cost percentage.

APPENDIX

Common Container Sizes

Canned Goods

Canned foods are packaged in standard-sized containers, but the actual amount of substance in a can may vary depending on the density of the food. Always check the label on a can or jar for the accurate net weight.

Can	Approximate Weight	Average Number of Cups per Can	Number of Cans per Case
No. 1/4 flat	4-3/4 oz.	1/2	24
No. 1/2 flat or 8 oz.	8 oz.	1	12 or 24
No. 1 Tall	12 to 16 oz.	2	12 or 24
No. 2	20 oz.	2-1/2	24
No. 2-1/2	28 oz.	3-1/2	24
No. 303	16 to 17 oz.	2	12 or 24
No. 5	3 lb. 8 oz.	5-1/2	12
No. 3 cylinder or 46 oz.	46 oz.	5-2/3	12
No. 10	6 lb. 10 oz.	13	3 or 6

Scoops

The number stamped on the scoop or disher indicates the *number of servings in a quart* (32 fl. oz.) of most foods. For example, a No. 16 scoop will give you 16 2-oz. servings from one quart of potato salad. The following measurements are approximate and will vary slightly depending on the food being measured.

	Level Measure	Fluid Ounces
6	2/3 cup	5-1/3
8	1/2	4
10	2/5 cup	3.2
12	1/3 cup	2-2/3
16	1/4 cup	2
20	3-1/5 Tbsp.	1-2/3
24	2-2/3 Tbsp.	1-1/3
30	2-1/5 Tbsp.	1
40	1-3/5 Tbsp.	.8
60	1 Tbsp.	1/2

Ladles

Ladles are used for portioning liquids such as sauce or soup. The handle of *each ladle is usually stamped with its fluid ounce capacity*.

Ladle Size	Portion of a Cup	Number per Quart
1 oz./30 ml	1/8 cup	32
2 oz./60 ml	1/4 cup	16
2-2/3 oz./80 ml	1/3 cup	12
4 oz./120 ml	1/2 cup	8
6 oz./180 ml	2/4 cup	5-1/3

APPROXIMATE TABLESPOONS PER OUNCE FOR SELECTED FOODS

ITEM	TBSP. PER OZ.	ITEM	TBSP. PER OZ.
Allspice, ground	5	Margarine	2
Baking Powder	3	Marjoram	10
Basil	8	Milk, dry	4
Celery Seed	6	Milk, whole liquid	2
Chili Powder	4	Mustard, dry	5
Cinnamon, ground	6	Mustard, prepared	4
Cloves, ground	5	Nutmeg, ground	6
Cloves, whole	6	Oil, salad	2
Cocoa Powder	4	Onion Salt	3
Coconut, grated	6	Oregano, ground	6
Coffee, ground	5	Paprika, ground	6
Corn Meal	3	Pepper, black ground	5
Cornstarch	3	Pepper, white ground	6
Corn Syrup	1 ½	Pickling Spice	8
Cream of Tartar	3	Poppy Seeds	6
Cumin	4	Poultry Seasoning	10
Curry Powder	8	Rosemary	10
Flour, all purpose	4	Sage, ground	8
Garlic Powder	4	Salt	2
Garlic Salt	3	Shortening	2
Ginger, ground	8	Soda, baking	3
Honey	1	Thyme	12
Mace	6	Vanilla Extract	2

RECIPE COSTING FORM

Menu Item _____ Date _____

Total Yield _____ Portion Size _____

INGREDIENT	QUANTITY	COST			RECIPE COST
		A.P. $	Yield %	E. P. $	

TOTAL RECIPE COST $ _____

Number of Portions _____

Cost per Portion $ _____

YIELD TESTING FORM

Food Item _____ Date _____

A.P. Price★ _____ A.P. Weight★ _____

Total A.P. Cost★ _____

Total Item Yield _____ Total Net Cost ❶ _____

Net Cost per Pound❷ _____ Percent of Increase❸ _____

By-products, Trim and Waste:

Item	Weight	$ Value	Total $ Value

Total Weight _____   Total Value $_____

Key to Calculations:

★   this information is taken from the invoice

❶   A.P. cost − by-product value = total net cost

❷   net cost ÷ item yeld = net cost per pound

❸   net per pound ÷ A.P. cost = % of increase
       % of increase × new purchase price = new net cost

COOKED YIELD TEST FORM

Item _____ Date _____

Net weight _____ Net Cost per Pound _____ Net Cost _____

Portion Size _____ Portions Served _____

Total Weight As-Served _____

Cooked Cost per Pound _____

Shrinkage _____ Percent of Shrinkage _____

Total Percent of Increase _____

DAILY PRODUCTION REPORT

Department _____   Date/Shift _____

Menu Item	Amount Ordered	Amount Produced	Number Sold	% of Sales	Leftovers

TOTAL _____

GLOSSARY

GLOSSARY

AS PURCHASED (A.P.) The condition and amount of an item as it is purchased or received from the supplier.

AS SERVED (A.S.) The weight, size or condition of a product as served or sold after processing or cooking.

BY-PRODUCT Anything that is made in the course of producing or preparing another item; it is secondary or incidental to the main product.

COST The price paid to acquire or produce an item.

COST PER PORTION The cost of one serving or saleable unit of a food; calculated as total recipe cost divided by number of portions.

COST OF GOODS SOLD The amount, at cost, of food items sold during a given period; calculated as food inventory at beginning of period plus food purchases during the period minus inventory at end of period.

COUNT The number of units or items.

CUTTING LOSS The unmeasurable but unavoidable weight lost from of a product during fabrication; caused by evaporation, blood loss, particles clinging to the knife, and so on.

DAILY PRODUCTION REPORT A list of the quantities of each item produced in the kitchen during a specific shift or day.

GLOSSARY

DIFFERENTIATE To distinguish a product or service from similar products or services, especially as perceived by customers.

DIRECT LABOR Labor that is used directly in the preparation of a food item.

EDIBLE PORTION (E.P.) The amount of a food item available for consumption after trimming or fabrication.

FABRICATED PRODUCT The item after trimming, boning, portioning, and so on.

FABRICATED YIELD PERCENTAGE The yield or edible portion of an item expressed as a percentage of the amount of the item purchased.

FOOD COST The cost of the foods and beverages that go directly into production of menu items.

FOOD COST PERCENTAGE The ratio of the cost of food served during a given period to the food sales dollars for that period.

GROSS PROFIT; GROSS MARGIN The difference between the cost of food served and the dollars taken in for food sales.

IMPERIAL SYSTEM A measurement system used in Great Britain and a few other countries; uses pounds and ounces for weights and pints for volume; imperial volumes are based on an imperial gallon that is equal to 4.546 liters or 1-1/5 U.S. gallons.

GLOSSARY

INVENTORY The counting and valuing of assets. *See* Perpetual Inventory *and* Physical Inventory.

INVOICE A document created by a seller to provide an itemized list of products and prices shipped to a buyer.

MENU A list of the foods and beverages, with prices, that are offered for sale by a food service establishment.

METRIC SYSTEM A measurement system based on units of 10; the gram is the basic unit of weight, the liter is the basic unit of capacity, and the meter is the basic unit of length.

PARSTOCK (PAR) The amount of stock necessary to cover operating needs between deliveries.

PERCEIVED VALUE The worth of an item from a customer's point of view.

PERCENTAGE OF INCREASE The increase in the value of a item expressed as a ratio of its purchase price (A.P. cost) to its net cost (after fabrication).

PERPETUAL INVENTORY An inventory maintained on an on-going basis by recording each receipt of an item and each issuance, use or sale of that item on individual cards or computer records.

GLOSSARY

PHYSICAL INVENTORY The listing and counting of all foods in the kitchen, storerooms and refrigerators.

PORTION One serving.

PORTION CONTROL The measurement of portions to ensure that the correct amount is being served each time.

PORTION COST The cost of one serving.

PORTION SIZE A specified portion amount.

PRIME COST The combination of food costs and direct labor costs.

PURCHASE SPECIFICATIONS Standard requirements established for procuring items from suppliers; also known as *specs*.

RECIPE COST The total cost of all ingredients in a standardized recipe.

RECIPE YIELD The count, weight or volume of food that a recipe will produce.

SHRINKAGE (1.) Loss attributed to dehydration, especially in meats, fish and some produce. (2.) Loss attributed to theft or some other unexplained event.

SPOILAGE Loss attributed to poor food handling.

GLOSSARY

STANDARDIZED RECIPE A set of instructions (formula) describing the way a particular establishment prepares a particular dish.

TRIM The part of the product removed when preparing an item for consumption.

UNIT A specified quantity, usually refers to the number or amount in a package; may be expressed as weight, volume, dozen, case, each, and so on.

UNIT COST The price paid to acquire one of the specified units.

U.S. SYSTEM A measurement system used in the United States; weight is measured in pounds and ounces, volume is measured in cups and gallons.

USABLE PORTION The portion of a fabricated product that has value; the amount being used in a recipe or the trim, fat or bones that has a resale value.

VOLUME The measurement of space or capacity; calculated as length × width × height; used to refer to the space occupied by a substance; expressed as teaspoons, cups, fluid ounces, gallons, liters, bushels, and so on.

WEIGHT The measurement of the mass or heaviness of a substance; expressed as ounces, pounds, grams, kilograms and so on.

GLOSSARY

YIELD (1.) The total created or the amount remaining after fabrication. (2.) The usable portion of a product.

YIELD PERCENTAGE; YIELD FACTOR The ratio of the usable amount to the amount purchased.

INDEX

INDEX

A
Abbreviations, 13, 73
As-Purchased (A.P.), 54
    costs, 36, 60-61
    quantities, 54, 63-64
As-Served (A.S.) *def.*,
    69

B
Budget-Based Pricing,
    115-16
By-products, 55, 65-66

C
Canned Goods, 122
Conversion Factor, 22-28
    and time, 31
Converting
    grams to ounces, 9-10
    recipes, 20, 23-28
Cost of Goods Sold, 83-86
    formula, 85
Cost per Portion, 45, 108
Count, 4, 6
    def., 6
    selling by, 7
Cutting Loss, 67

D
Daily Production Report,
    99-100
Direct Labor, 48-49, 111

E
Edible Portion (E.P.), 54, 60-61
    costs, 62
    quantity, 64

F
Factor Pricing, 109-10
FIFO, 97
Fluid Ounces, 5
Food Cost Percentage,
    86-89
    def., 87
    formula, 87
    menu pricing, 108, 109
Food Cost, 83-84, 94
    controlling, 94-95,
    97-99, 101

INDEX

def., 84
percent, 86

G
Grams, 9
  chart, 15
  to ounces, 9-10
Gross Profit Pricing, 114, 115

I
Imperial Measurements, 8
Indirect Labor, 48
Ingredient Price, 60
Inventory, 78-85
  def., 78
  form, 81
Invoices, 40-44, 97
Issuing
  of foodstuffs, 98

K
Kitchen Procedures, 94, 98-99

L
Ladles, 123
Labor Costs, 48-49, 65, 111
LIFO, 97

M
Market Price, 112
Measurement Systems, 7
  imperial, 8
  metric, 8
  U.S., 8
Measurements, 4
  chart, 14, 15
  imperial, 8
  metric, 8-12
  of ingredients, 4, 29
  of portions, 4
  U.S., 8
Menu
  and controlling food costs, 95
  and purchasing decisions, 74
  pricing, 108-18
Menu Planning, 84, 95

INDEX

Menu Prices, 108-18
    and food cost percent, 88
    and prime cost, 49, 111-12
    and profit, 114-15
    non-cost-based, 116-17
    psychological factors, 117-18
Metric Measurements, 8-12
    chart, 15

N
Net Cost, 68

P
Parstock, 95
Perceived Value Pricing, 112-13, 117
Percent of Sales, 101
Percent of Shrinkage, 71
Percentage of Increase, 68, 71
Perpetual Inventory, 78
    card, 79

Physical Inventory, 80
    form, 81
Pilferage, 83, 99-100
Portion Size, 20-21, 46
    converting, 26-28
    standards, 98-99
Price
    and menus, 108-18
    of ingredients, 60
    on invoice, 97
Prime Cost, 49, 111-12
    percent, 111-12
Produce Yields, 59
Profit
    and menu pricing, 114-15
Psychological Menu Pricing, 117-18
Purchase Specifications, 55, 59, 95
    chart, 96
Purchasing
    amounts, 62-64
    and food costs, 95
    and yield tests, 72-73
    decisions, 74
    form, 96
    of food, 36

INDEX

Q
Quantity
- to prepare, 20
- to purchase, 62-64

R
Receiving, 97
Recipe Conversions, 22-29
- and portion size, 26
- problems with, 30-32

Recipe Cost, 45-47, 61
- form, 46-47

S
Sales, 101
- percent of, 101

Scoops, 123
Selling Price, 108-18
Service, 101
Shrinkage
- controlling, 82-83
- *def.*, 71, 81
- of food, 71
- of inventory, 81-83
- percent of, 71

Specs, 95

Storing
- of foodstuffs, 97

T
Tablespoons per Ounce, 124
Temperature, 8
Theft, 83
Trim Loss, 55-56, 62

U-V
U.S. Measurements, 8
- *chart,* 14, 15

Unit Cost, 36-38, 60
- *def.,* 38

Volume, 4-6
- *chart,* 15
- *def.,* 4

W
Waste, 99
Weight, 4-6
- after cooking, 69-70
- *def.,* 4

Y
Yield

142

INDEX

def., 54
    form, 66-67
Yield Factor, 54
    applying, 60-64
    calculating, 55-57
    chart, 59
    of produce, 59
Yield Percent, 54, 55-57,
    58, 63
    chart, 59
Yield Tests
    cooked, 69-71
    for purchasing, 72-73
    with by-products, 65-68
    without by-products,
        55-57